NORTH AMERICAN INDIAN ECOLOGY

NORTH AMERICAN INDIAN ECOLOGY

by

J. Donald Hughes

With a new Foreword by the author

TEXAS WESTERN PRESS
The University of Texas at El Paso
1996

Second edition, 1996
© 1983 by Texas Western Press
the University of Texas at El Paso
El Paso, Texas 79968-0633

Library of Congress Catalog Card No. 94-061154
ISBN 0-87404-220-8

All rights reserved. No part of this book may be used in any manner without written permission from Texas Western Press, except in the case of brief quotations employed in reviews and similar critical works.

∞ Texas Western Press books are printed on acid-free paper, meeting the guidelines for permanence and durability of the Committee on Production Guidelines for Book Longevity of the Council on Library Resources.

*Dedicated to Ruth Murray Underhill:
friend, neighbor, and mentor.*

The signs of the dawn are seen in the east
And the breath of the new life is here . . .
Mother Earth is the first to be called to awake . . .
She moves, she awakes, she rises,
She feels the breath of the new-born Dawn.
The leaves and the grass stir;
All things move with the breath of the new day;
Everywhere life is renewed.

This is very mysterious;
We are speaking of something very sacred,
Although it happens every day.

— Tahirussawichi, Pawnee priest (Kurahus)

FOREWORD
TO THE
1995 SECOND EDITION

Today there is renewed interest by scholars and public alike in the ancestral and continuing relationship of the Native American Indians to the land and living beings on the North American continent. When "ecology" was a new word in the popular vocabulary, in the 1960s and 1970s, there were tendencies among non-Indians either to embrace a romantic view of the Indian and nature, or to reject as sentimental the idea that earlier Indians could have been ecologically sensitive. At the same time, Indians felt that others were misunderstanding or even exploiting a tradition that is their own, and is truly alive for them. There was reason for such a feeling. Some non-Indian authors were manufacturing "Indian" texts under pseudonyms such as "Chief Seattle," while others were writing books to prove that Indians never honored Mother Earth. In such a situation, many people did not know what to think, and their initial fascination with the subject consequently waned.

Now, however, the idea of Native American ecology is no longer new, no longer a fad. A body of respectable scholarly literature has appeared, written by Indians and non-Indians. Some descriptions of past thoughts and behaviors have been substantiated by sound research, while spurious documents and false claims have been examined and rejected. The picture that has survived this process is more realistic, and it shows us groups of real people who achieved sustainable ways of living on the North American earth. Those ways were not without hardships, and there were some missteps, but without denying the difficulties, it can still be affirmed that they walked in beauty. North Americans of every race and national origin are discovering that they can still learn from the Indians experience of nature on this continent, and they are doing so by turning to dependable sources and to Indian understandings of their own ancient heritage.

Indian people were here for thousands of years, became true inhabitants of every ecosystem, and are so still today. Where the past tense is used in this book, it is intended to affirm the past, not to suggest that the ideas and lifeways that are described have disappeared. Perhaps it should be explained why the term "American Indian" is preferred by many in modern times, but it fails to differentiate the descendants of the original inhabitants of the continent from millions of others who have been born in the same land since, and are therefore natives, too. "Amerindian" and "Amerind" smell of the lamp of scholarly anthropological invention. "Indian," despite its ambiguity, will have to do for now, until a better term emerges.

Finally, I would like to thank those thoughtful people who read the earlier edition of this book, made suggestions, and called mistakes to my attention. In particular, I have been encouraged by Indian people who read the book and told me that it was accurate and showed understanding. I also thank the editorial staff of the Texas Western Press for their careful and sensitive work, which has created a useful and beautiful book.

J. Donald Hughes
Denver, Colorado, 1995

TO THE READER

More than once I have made the journey to Zuñi Pueblo to watch the artist Alex Seowtewa at work painting murals on the inner walls of the old mission church. There emerging in bright colors are the masked dancers of the Zuñi festivals, shaking their hollow gourd rattles and wearing garlands of evergreen tree branches. And there too in the background is the landscape of Zuñi with its sacred mountains, changing with the seasons as the dances proceed through their yearly cycle. As one follows the mural around the church, one sees the snows of winter melt to be followed by the green grass and flowers of spring, the dryness and storms of summer, and the blue skies of autumn above yellowing leaves. Under the brushes and talented hands of this painter, "the Michelangelo of Zuñi," the interior of the adobe sanctuary becomes a celebration of the Indian perception of nature as sacred space, and time as a repeating cycle of wholeness. This scene is always a joyful surprise to me because it portrays the way two cultures can sometimes come together in understanding and create something of beauty that both can see.

It is too seldom this way. There was a time when the missionaries of the church destroyed such paintings wherever they found them, and there was a time when Indians in the Southwestern pueblos rose up and destroyed the churches. But today, in one place, love has won out at least for a little while, and the two ways of looking at the world have come together. The church has accepted the Indian spirit within, and Alex Seowtewa has freely made an inestimable gift. This is a parable of what I hope can happen in the dialogue between Native American Indians and other Americans concerning the many ecological crises that face us.

This book is a work of appreciation. It is an attempt to present from Indian sources the attitudes of Indians toward the natural environment and the practices that result from those attitudes. It is written with a sense of responsibility to many Indian friends and a sincere desire to tell the truth as I see it through my own research and experience. I realize that there are historians and anthropologists who will both agree and disagree with my conclusions. I have read their books and articles carefully, and I have tried to give constructive responses to their opinions in the following pages without any polemic or any attempt to argue with them point by point or by name. I dislike academic disputes, and will try not to be drawn into them. Wherever I have been informed by the work of these excellent modern scholars, I have tried to give full credit, but this book is not a review of the secondary literature on American Indian ecology; it is an attempt to listen to what Indians themselves have said and are saying about their relationship to the world of nature. Those interested in further reading in recent literature on the subject are referred to the bibliography.

This book would be incomplete without recognition of those fine Indian friends who have taught me not only through writings but by their spoken words and living presence. Among these are some who have served as co-teachers with me in classes at the University of Denver and have become my own teachers in so doing: Helen Peterson (Oglala Sioux), who deserves a special word of thanks for her energy and wisdom, Veronica Velarde Tiller (Jicarilla Apache), Patricia Locke (Chippewa-Sioux), Ray Rector (Cherokee), and, over a period of several years, Charles Cambridge (Navajo). I am grateful also for a mind-expanding class on Native American Religion taught at Iliff School of Theology by Vine Deloria, Jr. (Lakota Sioux).

In particular I wish to thank three individuals who have spoken with articulate wisdom about their world view, helping me to catch the vision that informs this book: Eddie Box, Sr., the Ute elder, who gave kind and eloquent talks to my classes in the Southern Ute Cultural Center in Ignacio, Colorado, and was our host at the Sun Dance; Alex Seowtewa, who came down from his scaffolding in the church at Zuñi and told us something of the vision that inspires his painting; and Emory Sekaquaptewa, a teacher and interpreter of his unique experience as a Hopi, who was willing to spend some hours talking with me about his people's view of the natural world, during the busy schedule of a conference at Sun Valley, Idaho.

Others have come to take part in my classes as speakers, to share their perspectives on many different aspects of American Indian life, among them: N. Scott Momaday (Kiowa), Richard Tallbull (Southern Cheyenne), John Fire Lame Deer (Lakota Sioux), Thomas Banyacya (Hopi), Clarence Acoya (Laguna Pueblo), Rachel Ashley (Lakota Sioux), Vernon Bellecourt (Chippewa), Gladys Benson (Mesquakie), Ruby Brave Eagle (Oglala Sioux), Gerald Clifford (Oglala Sioux), Richard Collins (Navajo), Ada Deer (Menominee), John Echohawk (Native American Rights Fund), Charles Emery (Cheyenne River Sioux), Joseph Enos (Papago), Roberta Hall (Winnebago), Vincent Harvier (Fort Yuma Apache), Farrell Howell (Pawnee), Ron Lewis (Cherokee), Willis and Regina Little Wolf (Chippewa and

Winnebago), Hartman H. Lomawaima (Hopi), Shirlene Nash (Apache), Phyllis Pierson (Mandan), John Redhouse (Navajo), Bernard Second (Mescalero Apache), Betty Lou and Darlene Silversmith (Navajo), Michael Taylor (Oglala Sioux), Charles Trimble (Lakota Sioux), Tillie Walker (Mandan), Walter Wetzel (Blackfeet), and Gerald Wilkinson (Cherokee).

As I have traveled to various places in Indian country, often with my students and sometimes to conferences, I have met many people of different tribes who were willing to talk and share their points of view, such as Robert Bennett (Oneida, former U. S. Commissioner of Indian Affairs), who was kind enough to write the introduction to an earlier book of mine, *American Indians in Colorado* (Boulder, Colorado: Pruett Press, 1977), Leonard Burch (Southern Ute Tribal Chairman), Roger Davis (Navajo), Wayne Sekaquaptewa (Hopi), Porter Timeche (Hopi), Hubert Velarde (Jicarilla Apache Tribal Chairman), and the Lelooska Family (Cherokee/Kwakiutl).

Some of my students have repaid my efforts by teaching me, too. Among these are Carol Bird Bear (Navajo), Patty Leah Harjo (Seneca-Seminole), Jack Lane (Standing Rock Sioux), and Robert E. Peck (Tlingit).

Also I owe a word of esteem to Clara Sue Kidwell, whom I have met and heard at several conferences, and whose concept of "ethnic science" seems to me to point the way to intercultural understanding in an important field.

Once Veronica Tiller's sister Bernice told me a delightful story about how Veronica's daughter Emily got her Indian name, which means "Flying Squirrel" in Apache. First she asked me to guess. I thought it might be because she was so quick and had such cute flashing black eyes. Pam, my wife, and I loved the little girl, who spent many days that summer at our house. "No," said Bernice, "it's because flying squirrels are red on one side and white on the other, and Emily is part Indian and part White." Well, I have had some "flying squirrels" in my family, too, and I have learned a lot from them over the years: an Apache brother-in-law named Johnny, and my sister Jane's three children: Michael, Suzanne, and Brian.

I am grateful to the Faculty Research Fund of the University Senate of the University of Denver for research grants used in the preparation of this book, to the four chairpersons of the Department of History, Allen D. Breck, John Livingston, Robert E. Roeder, and Charles P. Carlson, Jr., for their understanding and support, and to Mary R. Behner not only for immense help in the preparation of manuscripts but also for being one whose very presence in the right places made some remarkably good accidents happen – one thinks of words like "serendipity" and "synchronicity" – and enabled this book to take shape as it did.

If there are any I have forgotten to mention, and there undoubtedly are, I apologize. No one should assume that any of the people mentioned here necessarily agree with what I have written. They should not be held culpable for anything I say in this book, though whatever excellence there may be in these pages I am willing to ascribe to their influence.

Above all others I must mention in gratitude and respect Ruth Underhill, Professor Emeritus of Anthropology at the University of Denver and my neighbor and friend during fifteen years of her remarkable and creative life. She was who, never having really retired, came forth to help me start a program in American Indian History at the University of Denver and gave me support in essaying this present work.

J. Donald Hughes
Professor of History
University of Denver
November, 1993

CONTENTS

Chapter		Page
I	The Unspoiled Continent	1
II	The Sacred Universe	10
III	The Powerful Animals	23
IV	The Plant People	49
V	All Beings Share the Same Land	58
VI	The Gifts of Mother Earth	65
VII	The Wisdom of the Elders	78
VIII	Our People Covered the Land	95
IX	The Strangers' Ways	105
X	Indian Wisdom for Today	137
	Notes	144
	Bibliography	156
	Index	162

I

The Unspoiled Continent

LONG BEFORE the first European ship dropped anchor off the shores of the New World, the western continent was the home of the American Indians. They had lived here for twenty, thirty, forty thousand years. There was not a section of land unknown to some Indian tribe, and there was nowhere, from the slowly shifting arctic ice shelves to the blowing sand dunes of the Colorado Desert, where they did not go. Indians hunted buffalo on the plains and deer in the eastern forests. They planted corn in rich river bottomlands and near springs in the high desert. They caught salmon in the northwestern streams and set their boats on the Pacific waves in search of the great whales. Everywhere they went, they had learned to live with nature; to survive and indeed prosper in each kind of environment the vast land offered in seemingly infinite variety.

And they did all this without destroying, without polluting, without using up the living resources of the natural world. Somehow they had learned a secret that Europe had already lost, and which we seem to have lost now in America — the secret of how to live in harmony with Mother Earth, to use what she offers without hurting her; the secret of receiving gratefully the gifts of the Great Spirit.

When Indians alone cared for the American earth, this continent was clothed in a green robe of forests, unbroken grasslands, and useful

CHAPTER I

desert plants, filled with an abundance of wildlife. Changes have occurred since people with different attitudes have taken over. More than fifty years ago, an Omaha Indian elder expressed it this way:

> When I was a youth the country was very beautiful. Along the rivers were belts of timberland, where grew cottonwoods, maples, elms, ash, hickory and walnut trees, and many other kinds. Also there were various kinds of vines and shrubs. And under these grew many good herbs and beautiful flowering plants. In both the woodland and the prairie I could see the trails of many kinds of animals and hear the cheerful songs of birds of many kinds. When I walked abroad I could see many forms of life, beautiful living creatures of many kinds which Wakanda had placed here; and these were after their manner walking, flying, leaping, running, playing all about. But now the face of all the land is changed and sad. The living creatures are gone. I see the land desolate, and I suffer . . . loneliness.[1]

When the first European explorers coasted the shores, ascended the rivers, and trekked overland, they constantly remarked at the richness and variety of the new land: deep, fertile soil, flourishing woodlands, prairies full of high grass and myriads of flowers, and clear rivers of good water. The land teemed with the wildlife that sometimes did not even flee at the approach of the invaders. The waters had as many fish as the sky had birds, and Europeans had never seen so many birds — nor would they ever again. The marshlands thronged with them and flights of passenger pigeons darkened the sky for hours as they moved overhead like a living wind. The Indians had been hunting, fishing, and gathering in America not for centuries but for millenia, and there were still as many buffalo on the plains and salmon in the rivers as there could be. The land was not untouched, but it was unspoiled.

In fact, it was so unspoiled that the Europeans thought they were finding a "wilderness." Again and again that word appears in the explorers' and settlers' journals, and in all the European languages the word for wilderness means loneliness, a deserted territory, a land without human inhabitants. In English it is "wild-deer-ness," the place of wild beasts, not of men. To those European strangers it was either a threatening, ragged, untrodden tract empty of human life, or a Garden of Eden still as it was when the hand of the Creator rested, a sublime solitude unmarked by the axe, plow, or wheel. This was not the Indian view. Luther Standing Bear, reflecting on the way his people had looked at the world of nature, said:

> We did not think of the great open plains, the beautiful rolling hills, and winding streams with tangled growth, as 'wild.' Only to the white man was nature a 'wilderness' and only to him was the land 'infested' with 'wild' animals and 'savage' people. To us it was tame. Earth was bountiful and we were surrounded with the blessings of the Great Mystery.[2]

Then who were the Indians? — the newcomers asked. That they were all around, no European could deny. Every place the explorers went, they met Indians in hunting parties, farming towns, fishing villages, and further to the south, in cities with multi-storied dwellings and fantastic pyramids. How could a land thronging with inhabitants be a wilderness? Simply because it *looked* like a wilderness to the Europeans. It was unmarred, unexploited, to their eyes. They regarded the Indians as savages, wild denizens of a wild land, like a human species of predator in the forest. Or as the children of nature, living as noble, uncorrupted innocents in a state of grace, or at least ignorance of civilization. In either case, the Indians were thought to be few in number (European diseases, spreading from distant first contacts, often depopulated entire districts before they were entered by the explorers and settlers). And how could people who made so few changes in the land and forests actually be said to "occupy" or "own" the land?[3] Often the Europeans speak as if the new country had been uninhabited before they arrived.

But they were wrong. They could not have been more wrong. The condition of the New World as it met "the eyes of discovery" was a testimonial to the ecological wisdom of the Indians, both their attitudes and their ways of treating the natural environment. Nature flourished in the New World while the Old World was already deteriorating. Why? How was the North American continent preserved ecologically intact while it was in the hands of the Indians? What the Europeans saw before them was not a wilderness, an empty land. It was the artifact of a civilization whose relationship to the living world was perceived by the Indians in terms that Europeans would not grasp at all. If anyone asked Indians what they thought about animals, trees, and mountains, they answered by talking about the powerful spiritual beings that were those things. No European, whether he be Christian, rationalist, Jew, or deist, could possibly believe in ideas like those. They simply dismissed the Indians as believers in superstitution, worshippers of devils, or simple people who held naive primitive opinions that Europeans had once held, too, in the distant and unenlightened past. But the Indian attitudes — the Indian philosophy and religion, if those restrictive words can even be used to apply to the wholeness of Indian thought — enabled the Indians to live in and to change the American environment without seriously degrading it. Their very languages, which few Europeans bothered to learn, revealed a view of nature so foreign to that of the Europeans, and in many ways so far beyond it, that we are only beginning to appreciate it today. If all the resources of modern anthropology, psychology, and linguistics are only now piecing together the picture of the ecology and culture of Indians before Columbus, it is not surprising that the Europeans who first arrived did not understand what they were seeing.

It was not a wilderness — it was a community in nature of living beings, among whom the Indians formed a part, but not all. There were also animals, trees, plants, and rivers, and the Indians regarded themselves as relatives of these, not as their superiors. An Indian took pride not in making a mark on the land, but in leaving as few marks as possible: in walking through the forest without breaking branches, in building a fire that made as little smoke as possible, in killing one deer without disturbing the others.

Of course they made changes in their surroundings. All living things do; buffalos make wallows and bees build hives. Everywhere that people live, their activities have an effect on the natural environment. Mankind and human culture are agents of change in nature. Indians were no exception. Skilled, experienced Indian hunters killed moose in the north woods; they did so for thousands of years, long enough to exert a force of hunting selection on the moose population. They killed the slower ones, the less alert ones, more easily. So the moose of North America were different animals — faster, more alert perhaps — than they would have been if Indians had not been hunting them. Similar things happened to many other species in the forest ecosystems. The forest itself reflected the presence and character of its human inhabitants. Their land was not a wilderness, but a woodland park that had known expert hunters for millennia. Corn-growing Indians cleared the land, often by burning, and in some places in the South their planted fields stretched from one village to the next. Where they had established permanent towns, they tended to use up nearby trees for firewood: when Coronado came to Zuñi, he said the people had to go some distance away from the village to find junipers. But almost everything the Indians did kept them in balance with nature.[4]

For the Indians, living in careful balance with the natural environment was necessary to survival, since they lived so close to it and depended on it so completely. If they made serious mistakes in their treatment of nature, they felt the results right away; that is, they got immediate feedback. If they acted in ways that would destroy the balance of the natural communities where they found their food, clothing and shelter, then those communities would not provide for their needs any longer. Indians did not see this relationship as working in purely economic ways. Their actions were guided on every side by their view that nature is composed of a host of spirit persons who can talk to human beings and respond in a number of ways to the treatment they receive. they knew they had to be careful with those beings who shared the world with them because their lives were closely interlocked with them and they had to depend on them. And every part of their lives was involved in the relationship with nature.

For most modern urban people, our philosophy of life or how we think we see the world is one thing, and how we act in daily life is another. But for the Indians, life was all of one piece. How they perceived the natural environment, and how they treated it through the customary activities that were their ways of life, formed a consistent whole for each group.

So far we have spoken of "American Indians" as a single kind of human being, but they never thought of themselves in that way until they were forced to do so by European and non-Indian American attitudes and education. One of the most important things that can be said about Indians is that they are not all alike. In 1492 there were some 400 Indian languages spoken north of Mexico. They identified themselves as members of their own groups; for them, to say "the people" was to say Navajo (really "Dineh"), Kwakiutl, or Dakota, not "Indian." Ways of life, and therefore attitudes and practices varied from one area to another, and also between linguistic groups, tribes, communities, clans, families, and even individuals.

When a particular story or hunting method or agricultural practice or tribal ritual is mentioned in this book, we will take care to note the name of the group where it was known. Indian people are justly proud of their tribal traditions, and these must be treated with respect. Even

CHAPTER I

so, when we look at their relationship to the natural environment, there was a deeply rooted agreement in basic attitudes and in the general tendency of practices among all North American Indian groups in the centuries before Columbus. In spite of the evident differences in tribal ways of life and mythologies, there is an impressive underlying agreement in their expressions of reverence for the earth, kinship with all forms of life, and harmony with nature. So it is possible in a book like this one to speak of Indian concepts such as a sacred universe, powerful animals and trees, Mother Earth, learning about nature from the revered elders, and many others, because those ideas were held by every tribe. But we can never lose sight of the many-splendored variety of Indian languages, tribes, and traditions.

The diversity of their ways of life can be seen clearly in a few examples. In the forests of northeastern North America, a rich environment with many species of animals and plants, Algonquian and Iroquoian tribes lived by hunting, fishing and gathering. They planted corn, beans, squash, and tobacco, except in the colder north where the short growing season made agriculture difficult. Those near the marshes harvested the wild rice. Their houses, conical or domed, or large longhouses, were usually covered with bark. They wore soft leather clothing, traveled in canoes of birch or elm bark, and made snowshoes and sleds for winter travel. In their magnificent woodland environment the Indians lived, not as transient nomads but as permanent inhabitants, well aware of the natural balance in which they participated. Their way of life was perfectly suited to the forest, and neither scarred the land nor depleted the wildlife. When they burned or cleared land for planting, they did so carefully, and when they had lived long enough in one place, they would move their village to another place in their territory to let the earth recover.

The Great Plains, a land of grasses and grazing animals, provided the ecological setting for many different tribes. Here the great herds, millions upon millions of buffalo and antelope, elk and deer, provided the Indians with almost everything they needed for life. It was a climate of extremes of cold and heat, of drought and flood, of pleasant breezy days, lightning, prairie fires and tornados. It was also a place of beauty, carpeted with wildflowers in season and replete with birds whose very names tell us they sang on the wing: lark sparrow, lark bunting, horned lark. The Indian people of the high plains were hunters who followed the buffalo and lived in portable tepees. To the east, others had villages of large earthen or thatched lodges, and planted crops in the lowlands near the rivers, but they also traveled west on long seasonal hunting trips. These ways of life existed long before the domestic horses and rifles arrived with the Europeans. The Plains way of life was not easy, but it was well adapted to the surrounding ecosystem. The Indians themselves sensed the relationship and its beauty. The tepee fit harmoniously into its environment, shaped to let the wind flow around it without undue strain, and designed for ease of assembly and transportation. It was regarded as an image of the universe, its round base repeating the horizon's curve and its walls the sky; its total design reflecting the outline of a cottonwood leaf or the cross-section of a snowdrift. The Plains hunters expressed their thanks for the gifts of the buffalo by killing only as many as they needed and by using every part of the animals. Buffalo hides, well-tanned, gave them shelter. Buffalo hide sewn with buffalo sinew was the material of their clothes and moccasins. Buffalo chips were a source of fuel, since firewood was scarce on the Plains. Buffalo meat, sliced into thin strips and dried in the sun could be kept for long periods of time as "jerky," and then boiled to reconstitute a nourishing food. Bones, horns, hooves, and hair all had their uses in tools and ornaments. It was a way of life that celebrated the wholeness that a people achieves by interaction with the land where they live.

The dry American Southwest, with its variety of ecosystems and sharply delineated, colorful scenery dominated by bare rock forms, sculptured by erosion, was the natural environment for several distinct Indian groups. Along the Rio Grande and across that tablelands, the independent towns of the various Pueblo peoples still stand, made of the very stone and clay on which they were built, and looking from a distance like the formations of the earth itself. They raised maize and other crops with water from the rare but often violent downpours and from the few dependable springs and streams. Their cycle of ceremonies matched the return of the seasons of planting and harvest. Many of their chants ask the powers of the earth and sky for the life-sustaining rain so necessary in their arid environment. They felt that those powers both heard and answered, and they represented them by masked dancers whose rhythmic feet caressed the ground.

Along the Northwest Coast, where rain-laden clouds from the Pacific amply watered a giant forest of cedar, spruce, fir, and hemlock, a number of tribes lived on the abundance of trees and sea. The cedars, with their strong, straight-grained, easily worked wood, provided the material for a highly artistic Indian technology that produced large,

CHAPTER I

well built houses of boards and timbers, sleek dugout canoes, and the admirable wood carvings of masks and totem poles. Though the people of the Northwest did hunt animals in the forests and mountains, the major source of their wealth was found in the waters. Salmon made their way upstream in millions to spawn in the headwaters of the coastal rivers, and there were always enough of them and more for the nets and spears of the fisherfolk, but they were never wasted. Dried, smoked, or salted, they could be stored away for the rest of the year. In their seaworthy boats, hunters went out after seals, sea lions, sea otters, and even several species of whales including the spectacular orcas, or killer whales. They caught the salt water fish, too: sharks, herring, smelt, halibut and the eulachon (candlefish), whose copious oil filled the lamps on winter nights. Northwest Coast Indians survived by knowing their natural environment well and making direct use of its surpluses. It was a land of abundance, but that abundance was only available to those who had the necessary skills. They kept track of annual cycles, naming the months after the natural changes they observed. Their lives were closely involved in nature's rhythms, and they were conscious of this.

We have described only four ways of life among American Indians; examples from, as anthropologists say, four different cultural areas. There are many more, and no one should think that there are only four types of Indians. We could speak also of the Eskimos in the Arctic environments, the Paiutes of the Great Basin, the cornplanters of the southeastern Civilized Tribes, the people of the tropical Everglades, or the varied tribes of California. Each group learned to live within its own distinct environment by using the talents they had as men and women. Unfortunately, much scholarly writing on Indian "ecology" treats Indian cultures as if they were purely automatic and unconscious adaptations to the necessities of the environment. Such a view treats human beings as if they constituted another interesting species of wildlife. Indian conceptions of nature, and of their own place within nature, were determining factors in the impacts they made, or refrained from making, on nature. Indians related themselves to the natural world as human beings, who reasoned, remembered, made decisions, and honored traditions.

It would be hard to imagine any two ways of life that differ from each other more than those of the desert Pueblo planters and the fishers in the Northwest rain forests. But when we speak of their ecological values and their attitudes toward nature, there are striking similarities.

For all American Indians, ecology was not a separate subject or something to be concerned about only part of the time; it was involved in an entire way of life. For them, things that we would call ecology affected and were affected by everything they did. This is true of all people, of course, but American Indians seem to have been conscious of it in a special way. As Popovi Da, governor of San Ildefonso Pueblo, New Mexico, said, "The Indian's vital, organic attitude towards man's place within the framework of other living creatures has an impact on his actions, thinking, reasoning, judgment, and his ideas of enjoyment, as well as his education and government."[5] Viewing it in this manner, ecology is not in some separate compartment of life, but connected to everything else. It is not on the edge of life, but at its center. Native American Indians, with attitudes like this, lived gently with and maintained the health of the unspoiled American continent for thousands of years before Columbus.

A Crow (Absaroke) man gazes into the Black Canyon, part of his homeland. Photographer Edward Curtis, 1905.
COURTESY SMITHSONIAN INSTITUTION, NATIONAL ANTHROPOLOGICAL ARCHIVES.

Daisy Norris Gilham, a Blackfoot Indian woman, rides a horse that drags a travois, a method of transporting heavy loads across the plains. In earlier times, dogs were used to pull a travois. Photographer Roland W. Reed, 1915.

COURTESY SMITHSONIAN INSTITUTION NATIONAL ANTHROPOLOGICAL ARCHIVES.

II

The Sacred Universe

ANYONE WHO LOOKS at American Indian art with a sense of appreciation is impressed by the way in which it incorporates the images of nature. The designs represent clouds, the sun, moon, and stars, mountains, animals, birds, plants, insects, and the spirit beings that walk abroad in the world. Even the simplest decorated basket shows that the artist meant to relate his or her work to the whole universe. And a basket can be a microcosm, a mandala of the spirit life that Indians found both in nature and within themselves. Everything they made, whether it was painted pottery, weaving, embroidery, costumes, sand paintings, or petroglyphs, manifested their feeling for the many forms found in the natural world. A lodge of the Eastern plains, built of the same brown earth on which it stood, was made "round like the day and the sun and the path of the stars."[1] The startling shape and color of a Northwest carver's brilliantly painted eagle arrests us and says, "This man felt power in the eagle. He admired that great bird." Indians loved nature, not in any romantic sentimental way, but with an honest, respectful love born of daily contact. The Indian attitude toward nature was never merely utilitarian. The Pawnees sang of plants this way:

> "Spring is opening,
> I can smell the different perfumes
> of the white weeds used in the dance."[2]

They lived most of their lives in the out-of-doors, where they could look up to "behold the beauty of yonder moving black sky," to "behold the black clouds rolling through the sky," in the words of an Osage song. To think that they loved the changing moods of nature is not to read our own feelings back into the Indian experience; they themselves tell us how they felt in songs and prayers recorded long ago, such as this one of the Teton Sioux:

> "May the sun rise well
> May the earth appear
> Brightly shone upon
>
> May the moon rise well
> May the earth appear
> Brightly shone upon."[3]

The unmistakable Indian attitude toward nature is appreciation, varying from calm enjoyment to awestruck wonder. Indian poems, songs and descriptions are full of natural images that reflect a pure interest in environmental beauty. They liked to "listen to the song the needles make when the wind blows," according to Popovi Da, and "count the many shades of blue in the sky."[4]

Their attitude toward the natural world and their place within it was well expressed in the deservedly famous speech of Chief Seattle, of the Duwamish tribe, delivered before the governor of Washington Territory in 1853, at the new town that had been named in the Chief's honor. Seattle was fortunate in having a translator, Dr. Henry Smith, of considerable literary skill. Here is a portion of the speech:

> Our dead never forget the beautiful world that gave them being. They still love its winding rivers, its great mountains and its sequestered vales and verdant lined lakes and bays . . .
>
> Every part of this soil is sacred in the estimation of my people. Every hillside, every valley, every plain and grove, has been hallowed by some fond memory or some sad experience of my tribe. Even the rocks, which seem to lie dumb as they swelter in the sun along the silent sea shore, in solemn grandeur thrill with memories of past events connected with the fate of my people . . .[5]

The images that recur in the words of the great Indian orator are those of nature; "the return of the seasons," "the stars that never set," the grass, the trees. They recur even more powerfully in the songs and chants used in the sacred ceremonies. In Indian ritual poetry, some

CHAPTER II

natural objects and animals are not named directly, but referred to in short formulas that may remind us of the epithets of Homer by their cameo-like descriptions of natural characteristics. The Papagos, for example, may call the sun "the shining traveler," ground squirrels "stayers in houses," and the coyote "the woolly comrade"; while the Navajos can refer to the latter animal as "howler through the dawn."[6] Here are two lines from an Apache song celebrating a joyful union of the people with the source of all things:

> The sunbeams stream forward, dawn boys, with shimmering shoes, . . . On the beautiful mountains above, it is daylight.[7]

Deep appreciation of nature is not limited to a few tribes, nor does any tribe seem to lack it. Here is part of a Papago speech given at the time of purification after a pilgrimage to the sea to obtain salt:

> Then to the east they went, and, looking back,
> They saw the earth lie beautifully moist and finished.
> Then out flew Blue Jay magician;
> Soft feathers he pulled out and let them fall,
> Till earth was blue (with flowers).
> Then out flew Yellow Finch magician;
> Soft feathers he pulled out and let them fall,
> Till earth was yellow (with flowers).
> Thus it was fair, our year.[8]

A Zuñi rain prayer further illustrates the way in which sensitivity to natural beauty pervades ritual poetry:

> Yonder on all sides our fathers,
> Priests of the mossy mountains,
> All those whose sacred places are round about,
> Creatures of the open spaces
> You of the wooded places,
> We have passed you on your roads
> . . .
> You of the forest,
> You of the brush,
> All you who in divine wisdom,
> Stand here quietly,
> . . .
> You will go before.
> . . .

> We have given our plume wands human form,
> With the massed cloud wing
> Of the one who is our grandfather,
> The male turkey,
> With eagle's thin cloud wings,
> And with the striped cloud wings
> And massed cloud tails
> Of all the birds of summer.[9]

Navajo prayers constantly repeat the word *hozho*, which is environmental beauty, the happiness one experiences by being in harmony with nature. As the Navajo put it, "My surroundings everywhere shall be beautiful as I walk about; the Earth is beautiful." This is expressed in what is possibly the best known American Indian ritual poem, a song from the Navajo night chant:

> Oh you who dwell among the cliffs
> In the house made of dawn,
> House made of evening light,
> House made of dark cloud,
> House made of he-rain,
> House made of dark mist,
> House made of she-rain,
> House made of pollen,
> House made of grasshoppers,
> Where the dark mist curtains the doorway,
> The path to which is on the rainbow,
> Where the zigzag lightning stands high on top,
> Where the he-rain stands high on top,
> Oh, male divinity,
> With your moccasins of dark cloud, come to us.
> . . .
> In beauty I walk.
> With beauty before me, I walk.
> With beauty behind me, I walk.
> With beauty below me, I walk.
> With beauty above me, I walk.
> With beauty all around me, I walk.
> It is finished in beauty.[10]

The beauty referred to is both spiritual beauty and the pervading beauty of the natural world. And appreciation for it is expressed not only in words, but in "walking," that is, a way of living, a way of treating the

CHAPTER II

world. As Black Elk, a Sioux holy man, spoke in words addressed to Mother Earth, "Every step that we take upon You should be done in a sacred manner; each step should be as a prayer." 'Because You have made Your will known to us," he continued, "we will walk the path of life in holiness, bearing the love and knowledge of you in our hearts!"[11]

The attitude of Indians toward the natural environment was basically what we would call spiritual or religious, although religion for them was not separated from the rest of life. Their actions in respect to nature were in harmony with their view of the world as a sacred place, so if we wish to understand why they practiced conservation and avoided destructive exploitation, we will find that it is just as important to study their religion as it is to study their economy.

Indian languages had no word for "religion;" they expressed the idea by something like the Isleta Pueblo term "life-way" or 'life-need." To them, everything in their traditional way of life was sacred. For the Hopis, religion was simply "the Hopi way," including everything in life as Hopis saw and lived it. The Hopis spent about a third of their waking lives in ritual dances, prayers, songs, and preparation for ceremonials. But they did not see these activities as different, "Sunday" things. Though Indians would select special days for tribal celebrations, they felt that "Every dawn as it comes is a holy event, and every day is holy."[12]

So the Indians saw all their experiences with nature as having what we would call a spiritual dimension. The ethics that told them how to treat the environment was part of their religious world view. They would explain their attitudes toward nature in religious terms, and their religion was a religion of nature. It was simple in its general outlines and highly complex in its details, especially when tribal differences are taken into account. It had no systematic theology that can be subjected to the kind of rational analysis that the philosophers of non-Indian Western Civilization like to make. It had no "either/or." Some Indian religious ideas may seem to be contradictory at first glance to those educated in the European-American tradition. But if all Indian conceptions of nature are taken together, they will be seen to fit into a single, harmonious world view.

The Indians saw themselves as at one with nature. All their traditions agree on this. Nature is the larger whole of which mankind is only a part. People stand within the natural world, not separate from it; and are dependent on it, not dominant over it. All living things are one, and people are joined with birds and trees, predators and prey, rocks and rain in a vast, powerful, interrelationship. "The whole universe is enhanced with the same breath, rocks, trees, grass, earth, all animals, and men," said Intiwa, a Hopi.[13] "We are in one nest," was a Taos Pueblo saying concerning humans, animals and birds.[14] "That comfortable gap which we have left between ourselves and all other life on the planet, the Apache bridged in a stride."[15]

At the end of the Lakota Sioux cermony of the sacred pipe, all the participants would cry out, "We are related!" Joseph Epes Brown explains that these words do not only acknowledge "the relatedness of the immediate participating group. There is also an affirmation of the mysterious interrelatedness of all that is."[16]

The world, in Indian eyes, exists in intricate balance in all its parts, as male is balanced by female and the cardinal directions are in harmony with one another. Human beings must stay in harmony with it, and constantly strive to maintain the balance. The more powerful beings in the universe are not necessarily friendly or hostile to mankind, but rather indispensable parts of a carefully balanced whole, and therefore tend to sustain and preserve humanity along with everything else as long as the balance is not upset. This perception of nature is not so far from the ecological concept of the "balance of nature."

If people suffer, it is usually because they are out of harmony with nature, and though this is not always the fault of the individual, harmony must be restored. Every action in regard to nature must have its reciprocal action. The offerings made in many Indian ceremonies are not so much "sacrifices" as things given in exchange for other things taken or killed, to maintain the balance. A ceremony is one way in which people contribute to maintaining the world as it should be. Since mankind is related to the universe in reciprocity and balance, an act correctly performed should always obtain the appropriate response. A gift given compels a gift in return. A Papago, on offering tobacco or arrows, says "I give you this; now bring me luck."[17] In a Tewa myth, Spider Old Woman replied to a hunter who had given feathers to her, "We have to help you, because you never forget us; because you always take feathers out for us, we help you."[18] So Indians knew that human actions are a response to nature, but also that everything we do affects nature and calls forth a response. And the response is not impersonal.

They saw everything in nature as alive, not just animate, but fully alive in the way people are alive, conscious and sentient. The Zuñis, for example, called everything, whether it be a star, mountain, flower, eagle, or the earth itself, *ho'i*, a "living person." Some American

CHAPTER II

Indians were primarily hunters, gatherers, or fishers, and others were planters. But all of them looked on the natural environment as a world of spiritual reality. That is, the earth and the living creatures in it were not "things" to be used. They were living beings, personalities possessing power. So Indians did not feel themselves to be the only real persons in a world of things; in their experience, all creatures were alive with the same life that was in them, and trees and rivers, snakes, bluejays, and elk, reverberated with power and resonated with spirit.

When this way of looking at the world is explained to modern non-Indians, they often assume that Indian beliefs are an attempt to *explain* nature. Since Indians lacked concepts like atoms, cold fronts, the second law of thermodynamics, and germs, these people think, Indians thought up ideas like spirits, guardians, and the wisdom of animals to supply causes for what they saw happening around them. But this is not true. The Indian view of nature comes from deeper inside the human psyche than mere rational thought or intellectual curiosity, although Indians certainly had these too. But Indians regarded things in nature as spiritual beings, not because they were seeking some explantation for natural phenomena, but because human beings experience a spiritual resonance in nature. Indians feel a bear, a tree, a corn plant, or a mountain as a sentient presence that can hear and understand their words, and respond. A public ceremonial like a Pueblo Indian rain dance can be expected, if done properly, to set up the same kind of resonance with the clouds, so that the people are in harmony with the forces of nature, and receive what they need to live. All this is conceived as a process, not of bending nature to human will, but of subordinating the human will to natural rhythms. Sickness may be understood, for example, as the result of getting out of harmony with nature, and the healing process as one of re-establishing the harmony by removing impediments to it. It is quite understandable, then, that animals, fish, birds, and plants are invoked to aid medicine men and women.

Nature was to them a great, interrelated community including animals, plants, human beings, and some things that Americans of the Western European tradition would call physical objects on the one hand, and purely spirits on the other. No person, tribe, or species within the living unity of nature was seen as self-sufficient, human beings possibly least of all. The Indian did not define himself or herself as primarily an autonomous individual, but as a part of a whole; a member of the tribe, a living being like other living beings, a part of nature. Because of this deep kinship, Indians accorded to every form of life the right to live, perpetuate its species, and follow the way of its own being as a conscious fellow creature. Animals were treated with the same consideration and respect as human beings.

Mankind was not the master of life, but one of its many manifestations, related to all the other creatures. As Black Elk said, "With all beings and all things we shall be as relatives."[19] To Indians, "Man is not lord of the universe. The forests and the fields have not been given him to despoil. He is equal in the world with the rabbit and the deer and the young corn plant."[20] Nature was not some European feudal fief with mankind as steward or subduer, but a primitive democracy in which every creature had its place, with privileges and duties to the others. In fact, the Indian view of the human role in nature is almost the reverse of the Western European understanding of "dominion." Greater power resides in natural forces and spiritual beings than in the hands of mankind. Human beings cannot dominate the natural world, for it is vastly more powerful and enduring than mankind.

> The Cheyenne in no sense believes that he can control nature. Although his environment is hard and life is precarious, he sees it as a good environment. It is one, however, with which he must keep himself in close tune through careful and tight self-control. . . . Man must fit himself to the conditions of environmental organization and functioning.[21]

Animals and plants can be addressed in prayer because, in all of nature, they are powers closest to human beings and may be willing to help them, provided they are approached in the proper manner. They are like human beings — and may even be human beings of another sort, or in disguise, or on another level of existence — but they are more mysterious, more holy, closer to the sources of power. They must be given at least the same respect and reciprocal fairness one would accord to a member of one's own tribe.

So everything in nature was powerful, able to help or harm. Mankind depended on the other beings for life, and they depended on mankind to maintain the proper balance. Living things must not be hurt or killed needlessly. If any species were totally destroyed or driven away from an area, it would leave a terrible gap that could not be filled unless it were to return. Mankind should hold a reciprocating, mutually beneficial relationship with each type of being. And as a result, all

CHAPTER II

Indian groups were very careful about how they treated animals, plants, and every other part of nature. They developed practices, differing in detail from place to place, that tended to conserve living creatures and preserve the balance of nature within their own living space.

American Indian ethics in regard to nature is, therefore, protective and life-preserving. It is a combination of reverence for life and affirmation to life, of which Albert Schweitzer would doubtless have approved. To Indians, the earth was a reality, not illusion, and it was loved, not callously exploited. The Hopis prayed for the welfare of all living things. Nothing could be killed except out of necessity. All tribes had similar ideas, feeling that every creature must be treated with care, never injured. The real "people" living in the world are not humans alone, but all spirit beings.

This viewing of the world in a sacred perspective was therefore, the caring for every aspect of the natural environment: "the wingeds, the two-leggeds, and the four-leggeds, are really the gifts of *Wakan-Tanka*. They are all *wakan* and should be treated as such."[22] This saying of Black Elk's uses two Dakota words for concepts that are fundamental to understanding how Indians related to the environment. The first is *wakan*, "power," the sacred power that permeates all natural forms and movements. The Indian's world was full of power, and of beings who had power, or were power. It would not be wise to attempt to define the exact relationships between these powers, or to make too fine a destinction between personal and impersonal power. The power that animates the universe and gives it regularity was described in one aspect as an impersonal force, or in another aspect as a personal deity, but these ideas were not opposed; they were two ends of a continuum. The Iroquois experienced an invisible force which they call *orenda*; other tribes had other names for a mysterious power that might often manifest itself in natural phenomena. The *wakonda* of Siouan-speaking tribes had its counterpart in *maxpe* of the Crows, the Hidatsa *xupa*, and the Algonquian *manito*, all standing for the power perceived in nature. More personalized was *Tirawa* of the Pawnees, a supreme being who revealed himself through nature, or the "Wise One Above," *Heamma wihio* of the Cheyennes, whose emblems were the sun and the spider that spins from itself.

The second word used by Black Elk is *Wakan-Tanka*, his people's name for the Great Spirit or "Great Mystery." The sense of a "Master of life," one spirit who breathed in everything and included all other spirits, existed in virtually every tribe.²³ Names in the different Indian languages show how they understood the Great Spirit. The Apache supreme being was called Life Giver. The Algonquian *Manitou* and the Cherokee *Esaugetuh Emissee* mean "Giver of Breath." Others, like the Papagos, used a word meaning "Earthmaker;" the Crow word *Ahbadt-dadt-deah* signifies "The One Who Made All Things." There was the concept of a lofty creator being who made everything, like *Alquntam* of the Bella Coola, "from whom come, and to whom belong, all myths,"²⁴ or the Haidas' *Sins Sganagwai*, "Power of the Shining Heavens," who was believed to give power to all things in nature, and of whom it could be said, "whatever one thinks, he knows."²⁵ Although some tribes did not necessarily regard the Great Spirit and the Creator as the same, such an identification was usual. Black Elk gave voice to what was no doubt the feeling of most Indians when he prayed,

> O Father and Grandfather *Wakan-Tanka*, You are the source and end of everything. My Father *Wakan-Tanka*, You are the one who watches over and sustains all life. O my Grandmother, You are the earthly source of all existence.²⁶

The world and all the good things in it were seen as the gifts of the Makers, to be received and used with thankfulness and reverent care. In this prayer, as is so often true in Indian expressions, the Great Spirit is addressed with terms of relationship that are both masculine and feminine. Tribes like the Utes and Zuñi used words meaning "The Great He-She." The Tewa creators, the Corn Mothers, were clearly female. The result of all these ideas for Indians was that everything in the natural environment was seen as a gift of the Great Spirit. As Black Hawk said, "I never take a drink of water from a spring without being mindful of His goodness."²⁷

But the Great Spirit was not alone in the world. The Indians recognized that nature is complicated, not simple, and there are things that are hard to understand. Why is the land so often steep and rocky along our trail? Why are rivers so crooked? Why do we have to die? Stories are told of a devious counterpart of the Creator, The Trickster, who was responsible for many of the paradoxes in nature. Although he had been the cause of much trouble, he was not really unfriendly or the enemy of mankind. It was just that he had a way of turning things upside down, of being lazy or contrary, of playing jokes on others. Sometimes his tricks were blessings in disguise. He, too, had many names: Warty, Flint, Rabbit, Bluejay, Coyote, etc. According to the Navajo, when the world was created, Coyote was given a blanket full of stars to place in the sky, with instructions to put them carefully in rows, equally spaced. Had he followed the Creator's orders, the sky might now look like a larger version of the blue field of the American flag. But after putting a few stars in place, Coyote decided that the whole job was too much work, so muttering the Navajo equivalent of "the heck with it," he flipped the blanket and sent the stars spinning across the sky into the magnificent disarray we now behold.[28] But who could find a way across the land and sea at night if all the stars looked the same? Coyote's pranks were sometimes good, sometimes bad. The Zuñis said he made maize edible. He was also responsible for the fact that stones will not float in water, and that people die. But if people did not die, the great circle of human life could not be completed, and then how could babies be born? So the Trickster, the dark side of the Creator, is necessary to complete the sacred universe.

Still another great power was Mother Earth herself, or Grandmother Earth, a generous being who supported mankind, providing fruits, roots, fish, and animals. In the important Apache coming of age ceremony, the young woman is seen as becoming one with the fruitful Earth Goddess. Earth is a basic reality. A Mandan song repeats, "earth always endures," and a Teton song might be phrased "Old men, you said earth only endures; you spoke truly, you are right."[29] Such was the death chant of White Antelope at Sand Creek: "Nothing lives long except the earth and the mountains."[30] Mother Earth was respected as the continuing source of life for all living creatures, giving birth and sustenance, as more of Black Elk's words explain:

> We are of earth, and belong to You. O Mother Earth from who we receive our food, You care for our growth as do our own mothers. Every step that we take upon You should be done in a sacred manner; each step should be as a prayer.[31]

The practical result of this perspective was care for the natural world. Since the earth herself was conceived of as being alive, an aspect of that care was to refrain from harming the earth. For this reason, Indians often objected to frontier miners who dug holes in the ground, or farmers who plowed, thus tearing the breast of Mother Earth. Indian farmers used a digging stick, an implement that symbolized the natural process of fertilization.

They were taught their useful arts by legendary givers of culture, for the spirits were the first farmers, hunters, singers, runners, weavers, and players of games. The Papagos' Elder Brother or Morning Star taught the art of making a wine from the fruit of the saguaro cactus. Some culture heroes were animals; many groups say, for example, the Mockingbird taught languages and songs and Spider Woman taught weaving. Among agricultural people, the giver of maize had a high standing. For the Navajos, it was Talking God who gave maize to Whiteshell Woman and her sister, Turquoise Woman. Other agricultural deities include the Zuñi Corn Maidens; Iyatiku, the Keres creator who is also goddess of maize; Muingwu, the Hopi god of vegetation; and Masau'u, the great god who met the Hopi and showed them his ability to grow crops to maturity in a single day. Poshaiangkia, the Zuñi culture hero sometimes given the inaccurate name "Montezuma," and who had the Beast Gods as his warriors, was traditionally the teacher both of agriculture and of ceremonies.

Great natural phenomena were regarded as particularly powerful spirits. The Sun, a powerful being, was sometimes closely identified with the Master of Breath, and as the evident quickener of life, was an important deity for all tribes. Sunrise was the greatest daily event, and the annual march of the Sun from north to south and back again was seen to control the seasons, the growth of crops and the habits of wild creatures. Prayers were said at dawn and ceremonies performed at the solstices. Children were presented to the Sun, and he was often regarded as the father of heroes.

Other powers of the sky were personified: often Sky himself, the Thunder People, Rainbows, and the Stars. Shotuknangu was the Hopi god of the sky, stars, and lightning. Other superhuman entities were the Moon, the Winds, and Fire. There were countless beings abroad in the world; water monsters, little people, giants; all dwellers of the forest who expected gifts, and who might cause or cure diseases or psychoses. Windigo, the cannibal spirit, was especially feared. The Iroquois false face curing societies attempted to use the power of the forest spirits for the good of tribal members. Masks representing these beings were carved directly in the wood of standing trees, and used in ceremonies.

As we have seen, animals and plants were seen as spirits, too. The eagle soared so high in the sky that he was identified with the sun. Like human beings, the other creatures were believed to worship the mysterious power in which they shared. The leaves of cottonwood trees

CHAPTER II

rustling in the wind were believed to be their voices praying to the Great Spirit who gave them the power to stand upright.

All the outward forms seen by Indians in the natural environment concealed personality and power which might be invoked. So they were constantly speaking to those manifested inner realities in words that, they trusted, were understood. A daily morning prayer was usually addressed to the sun, as in this Kwakiutl example: "Look at me Chief, that nothing evil may happen to me this day, made by you as you please, Great-Walking-To-and-Fro-All-Over-the-World, Chief."[32] As the careful man passed beside a steep mountain, he would speak to it, "Please make yourself firm."[33] Migrating birds were asked to take sickness far away. A Clayoquot Indian sang to a rough sea,

> Breakers, roll more easily.
> Don't break so high
> Become quiet.[34]

There were traditional prayers to use when nearing a waterfall, first seeing a lark, or glimpsing almost any animal of land, sea, or air. Walking about or gliding in his canoe, a traditional Indian would be holding constant conversation with the sacred universe.

Ecologists in recent years have been trying to get people to think of the world in ecological terms: to see everything as connected to everything else, to see ourselves not as the rulers of the earth but as fellow citizens with all other forms of life, and to see the earth as a biosphere in which natural systems operate endlessly to recycle water, oxygen, nutrients and energy among living and non-living parts of the environment. But American Indians would have recognized these ideas as soon as they were explained to them. Their philosophy was already ecological. When they wanted to make a picture of the universe, they drew a great endless circle, perhaps adding the lines of the four directions inside. To them, everything was connected, everything partook of the roundness, everything shared the same life.

The Sacred Universe. A Cheyenne Indian, named Above Bear, prays with a sacred pipe which is the symbol of the essential unity between human beings and the natural environment. Photo taken in the Southern Cheyenne Colony, Oklahoma, by George Bird Grinnell.

COURTESY MUSEUM OF THE AMERICAN INDIAN, HEYE FOUNDATION.

The Sacred Universe. The Buffalo Dance at San Ildefonso pueblo, New Mexico, represents several species of wild animals. Photographer Matilda Coxe Stevenson, 1909.
    COURTESY SMITHSONIAN INSTITUTION NATIONAL ANTHROPOLOGICAL ARCHIVES.

The Sacred Universe. Deer dancers, in the Medicine Ceremony of the Arikara tribe, circle the sacred cedar tree which is ritually felled in order that it may serve as the focal point of the dance. Photographer Edward S. Curtis, 1908.

COURTESY SMITHSONIAN INSTITUTION, NATIONAL ANTHROPOLOGICAL ARCHIVES.

The Sacred Universe. The Shu maakwe Society Dance, sometimes called the Shumai koli Dance, is performed at Zuñi by circling the sacred ponderosa pine tree. Photographer Matilda Coxe Stevenson. COURTESY SMITHSONIAN INSTITUTION NATIONAL ANTHROPOLOGICAL ARCHIVES.

III

The Powerful Animals

"WHEN THE HUNTER goes into the woods to hunt . . . and he sees a black bear he shoots it. As soon as he has killed it, then the hunter goes to where the bear that he has shot lies dead and stands by its side and says, 'Thank you, friend, that we met. I did not mean to do you any harm, friend.' . . ."[1] In these words of an Indian from the British Columbia coast can be sensed an attitude that is seldom if ever voiced among non-Indian hunters today. The Indian hunter felt sure that the animals he hunted were powerful beings who could hear and understand him, so he said a prayer of thanks and apology, explaining that he and his family needed to have the food and fur that the bear had provided.

Hunting was an activity familiar to every North American Indian tribe. Some hunted as their major way of getting food to eat, along with fishing and gathering wild plants. But the agricultural tribes knew hunting, too. Their ancestors had been hunters before they learned to plant maize and other crops, and that tradition was kept alive as agriculturalists supplemented their diet with meat from wild animals. Those same animals played a role in their ceremonial life that was much more significant than might be expected among farmers.

We have seen that Indians believed animals and human beings are closely related. Animals, including fish, birds, and even insects like the woodworm, were believed to have been created at the same time as human beings, and in similar form. In Navajo stories, for example, animals and people emerged together from the underworld. Animals had the ability to shed their coats of fur and appear as human beings, just as some human beings learned the art of assuming animal form. In fact, animals were seen as beings who put on their animal forms much as a person might don a mask or costume. At the creation, these beings had each chosen a specific animal form, according to their own temperaments and preferences. They continued to return to the world

CHAPTER III

for the benefit of human beings, willingly sacrificing their beings, as long as they were treated with care and honor. The rest of the time, they lived in their own "villages" under the sea, or underground. According to Northwest Coast Indians, porcupines had their village in a giant spruce tree, the beavers in a lake, the mountain goats on top of a large, flattopped mountain, and the salmon far off to the west on the bottom of the sea. When they were killed, they did not die, but simply discarded their outer "masks" and returned to their villages, from which they could return again, having put on a similar "mask." But those who had been mistreated might never return. All animals, fish, and birds were believed to be related to each other, able to understand one another's languages and the languages of human beings. Once, people could talk with the animals, and a few individuals, it was held, still received the ability to understand them. As Luther Standing Bear puts it,

> For the animal and bird world there existed a brotherly feeling that kept the Lakota safe among them and so close did some of the Lakotas come to their feathered and furred friends that in true brotherhood they spoke a common tongue.[2]

Animal songs were sung, worded as if spoken by the animals. The Navajos had a song for practically every known creature; the owl, for example: "I do not want the night to end."[3] Such songs had magical power to be heard by the creatures, and to charm them. Some gifted Indians could interpret the songs and cries of birds; they said that ravens could warn them of changes in the weather, movements of animals, and misfortune. Stories recount the visit of a hero to a tribe of animals who live in a village like people, where he may have watched the animals enacting their own dances and ceremonies.[4] The tales are also told of children adopted and raised by deer or antelope, and of a boy who came to prefer the company of eagles to people, who finally flew away into the sky, transformed into an eagle.

Marriages between people and animals in human form were common in legend. Such things were not necessarily expected to happen in the experience of the hearers, but they were taken seriously as descriptions of the remote past and reveal the Indian belief in animal-human kinship. Slain bears, mountain lions, and eagles were adopted into the tribe. Many dead people became owls, in whose cries could be heard their names, according to the tribes of the northern Pacific Coast, but dead hunters went among the wolves, and sea hunters to the killer whales. None of these species were supposed to be killed, but twins were believed to be related to salmon, and went to live in salmon-town when they died. Among the Indians of the Great Plains, the most significant of all animals were the buffalo, whose power was evident in their strength, and whose presence was necessary to the people's very survival. They were always conscious of their dependence on the buffalo, whom they regarded as closest of all animals to human beings. The buffalo evidently had a religion, for they were seen to purify their "children" by washing them, and made "offerings" of hair when they rubbed against trees.

Animals thus were regarded as closely related to human beings, but also as powerful spirits with mysterious, separate lives of their own, not lower, but if anything on a higher plane than human beings and able to help or hurt with their power. The killing of animals was not to be undertaken carelessly. So the hunter practiced an art, at every step, with sacred ritual designed to preserve the balanced relationship between man and nature. Hunting was "not war upon the animals, not a slaughter for food or profit, but a *holy occupation*."[5] Life, Indians believed, is fostered and prolonged by right living, mindful of the individual's kinship both with other members of the tribe and the creatures with which they shared the earth. They considered hunting to be a spiritual encounter between two conscious beings who stood in reciprocal relationship to one another, a relationship that operated through ritual. As Irving Goldman explains it,

> The encounter between the chiefly hunter and his prey seems to involve a vital interchange. The animal yields its life for the welfare of the hunter and of his community. The hunter dedicates himself in turn to the rituals of maintaining the continuity of the life cycle for all. In many instances the hunter does submit to ritual preparations for the hunt not, as it is often thought, to insure his success, but as the correct and courteous way of meeting the animal who is going to make him a gift of its life. Imputing to the animal a willingness to die shows no intent to gloss over the aggressiveness in the act of killing.[6]

Rituals were necessary, because Indians believed that the creatures could not be killed unless they offered themselves willingly for food and clothing, so that people could continue to live and thrive upon the earth. Animals revealed their whereabouts to a good hunter, and

CHAPTER III

allowed themselves to be taken willingly, but only if they were treated with consideration and respect. The hunter and his victim both understood their roles. The Good Hunter was celebrated in many legends as a model worthy of imitation. He was one who killed only when he needed food, left gifts in the forest for the animals, and treated them always with kindness and patience. In return, the animals helped him in the hunt and might even revive him after he had been killed in war, or so it was said.

Hunting ritual was intended to entreat the animals not to get angry if they were killed, but to allow themselves to be taken for the use of their human kinspeople. It was believed that spirits of slain animals did not die, but went to their own country to report on their treatment by human hunters, and either to return in gratitude to be hunted again, and encourage others to do so, or to send sickness and accident, depending on how the hunters had behaved toward them. The Micmac Indians reported that one time when too many moose had been killed wastefully, all the remaining moose left the vicinity.[7] The Indians felt that they had to consider the feelings of the animals they killed, and those of the survivors as well. In addition, those already killed had not really died, but might not return. Part of the purpose of the hunt was to get the animals to give their abilities and good qualities to the hunter who killed them. A Kwakiutl prayer to a beaver exemplifies this idea:

> Welcome, friend Throwing-Down-in-One-Day, for you have agreed to come to me. I want to catch you because I wish you to give me your ability to work, that I may be like you; for there is no work you cannot do, you Throw-Down-in-One-Day, you Tree-Feller, you Owner-of-Weather, and also that no evil befall me in what I am doing, friend.[8]

The same principle applied in the hunting of bears; since the fearless, ferocious characteristics of the grizzly were those desired in a noble or chief, it was felt in many tribes that only leading men should be allowed to hunt them and take their qualities. A Kwakiutl could pray to a slain black bear:

> Thank you, friend, that you did not make me walk about in vain. Now you have come to take mercy on me so that I may inherit your power of getting easily with your hands the salmon that you catch.[9]

Note that in these examples the words "beaver" and "bear" are avoided. The hunter believed that the animal could hear and might flee if the specific names were spoken, so a name of honor was used instead.

Prayers, songs, offerings, and observance of carefully framed rules of behavior were part of hunting in every tribe. They were highly developed because they had been elaborated for thousands of years. A tribe whose traditions can be traced back to ancestors who were hunters in the forests of western Canada, the Navajos, had a series of hunting rituals of staggering complexity. There was a separate "way" to learn not just for each kind of animal, but for each method of hunting each animal: the Wolf Way to drive deer, the Stalking Way to use a deer disguise, and the Big Snake Way to ambush deer, to list no more than three examples.

Hunting rituals include those that prepare an individual to be a hunter, those used before going out to seek the animal, those used during the hunt up to the time of the kill, and those used afterwards, on the way home and in the camp or village. First of all, in many tribes a young man who wanted to become a hunter was expected not only to learn his art from an accomplished elder, but also to seek the approval of the animal spirits. A hunter's skill might have been the result of practice, but it would have lacked any effect without the special power that came as a gift, quite likely in a dream or vision. The vision quest was a disciplined means of gaining help from transhuman sources of power. After preparations including abstinence from food and sex, the individual had a personal, mystical experience in which a certain spirit appeared and granted power. Usually the spirit proved to be that of an animal, and the hunter would seek help from that particular animal spirit for the rest of his life. The hunter trusted his guardian spirit, and sought the animals he had been given the skill to hunt. He sang the songs he had learned in the vision to get power for hunting.

Each time there was a hunt, there were preparations to be made. Before going after his quarry, the hunter almost always went through a purification of bathing, rubbing with plants, fasting and continence. His wife might be warned to move easily and quietly while he was away, obeying numerous taboos, since her actions were believed to influence those of the animals.

Preparation for a hunt was a time of careful rituals to invoke hunting power, and a time of taboos to avoid contact with contrary power. There were songs to sing, prayers to be said, and offerings to be made of meat and other things. When Northwestern Indians hunted for mountain goats, the head of one might be brought near the fire where it could be warmed, presented with offerings, painted the "way a mountain goat painted his face when he killed a man."[10] In many tribes across the continent, bear hunting involved more ceremonial than that for any other animal, although the details varied from place to place. The hunter might subject himself to an unusually severe purification. Putting red paint and eagle-down on a bear's head, he would pray, "I am your friend, I am poor and come to you."[11] Pueblo Indians made prayer-stick offerings at animal shrines, with prayers of apology. Talismans of the beast gods were taken out, ritually "fed," and asked to lead animals to the hunters. Feathers, hides, claws, shells, other animal parts, and even stones shaped like animals provided power in ceremonies, because a part could stand for the whole, and an object that resembled something could act for it. This is the reason a talisman could help both in the hunt and in healing; in resembling the animal, it partook of the quality of the animal and therefore its spirit power. A sacred bundle consisted of a series of talismans taken from nature and embodying power from nature. Tobacco might be "fed" to the hunting talismans. In virtually every case, the hunter was a man, and women were especially avoided by him since their power was of a different kind and might render his weapons ineffective. The weapons themselves were treated with honor and kept carefully; they would be buried with the dead hunter so that his spirit might use their spirits even after death.

Hunting rituals were often not just an individual matter, and preparation for the hunt could involve much ceremonial. Among the Pueblos, permission might have to be asked from designated individuals, and it was given in a formal speech of forgiveness for the killing that would be done out of necessity, not cruelty. Papagos also made a ritual oration before the hunt, in which again care was taken to avoid using the real names of the animals in order to spare their feelings. A whole tribe often joined in ceremonials for important seasonal events, such as the time when bears emerged from hibernation, or at the beginning of the hunt in the fall. The bear was given a major role in these celebrations in the eastern forests. A powerful animal with a somewhat manlike appearance, the bear received an extra degree of ceremonial treatment, and was addressed with honorific titles, even terms of relationships, such as "my paternal uncle."[12] Sometimes the relationship claimed was very specific, such as the Tlingit who prayed to the bear: "My father's brother-in-law, have pity on me. Let me be in luck."[13]

Among the Plains Indians, buffalo were honored in song, art, and worship, and the entire tribe participated in preparations for the buffalo hunt. It was preceded by offerings, fasting, prayers, the building of altars, and the smoking of pipes; and the hunt itself was often conducted as the most awesome of ceremonies, begun with panoplied processions and surrounded by ritual prohibitions. Unlike the game of the forests, the buffalo were herd animals, and the buffalo hunt was usually a communal activity. Propitiation of the buffalo was, therefore, not a matter of apology by the individual hunter to the individual animal, but was done in tribal ceremonies augmented by visions received by individuals to benefit the tribe. Two of the Sacred Arrows of the Cheyennes were considered to have power to bring the buffalo. An Osage could sing a buffalo song with the same purpose:

> Look you, my grandfathers rise,
> They of the shaggy manes, rise quickly;
>
> Look you, my grandfathers rise,
> They of the curved horns, rise quickly;
>
> Look you, my grandfathers rise,
> They of the humped shoulders, rise quickly;
>
> Look you, my grandfathers rise,
> They whose tails curl in anger, rise quickly;
>
> Look you, my grandfathers rise,
> They, the four-legged ones, rise quickly;
>
> Look you, my grandfathers rise,
> They who paw the earth in anger, rise quickly.[14]

Ritual continued throughout the days of the hunt, having as possible purposes both the attraction of the animals and the stabilization of a psychological attitude of respect and apology on the part of the hunter which protected the fauna by preventing waste and overkill. It was considered especially dangerous to laugh or use bad language when hunting. The first animal seen of the chosen species was usually let go, with a speech telling it of the need of the hunter and his people, and asking it to tell the others of its "tribe" to come and offer themselves to be killed. Medicine men with special power for the particular animals could go along on the hunt. A Papago medicine man traveling with

CHAPTER III

hunters would go out into the night to sense which creatures wanted to offer themselves to be killed. The Plains Indians spoke of their hunt not as "driving" the buffalo but as "leading them;" not "chasing," but "calling." Certain men, often of certain societies such as the Beaver Men among the Blackfeet, or the Cheyenne Elk Society, were given the responsibility and known to possess the power of charming the buffalo into a corral or over a cliff. They did this by putting on buffalo robes and crawling on all fours among the herd, imitating the movements and making the sounds, perhaps through a megaphone of hide or bark, of a cow who was searching for a lost calf.

When a living creature of any kind was taken by the hunter, he had to make an apology to it as a spiritual being, thanking it for the sacrifice it had made, and explaining the need that had made the killing necessary. Even when snares or deadfalls were set for small animals, ritual offerings would be left nearby for the animals' spirits. When a Papago hunter had killed a deer, he turned its head in the right direction and apologized: "I have killed you because I need food. Do not be angry."[15] His Navajo counterpart might chant a much more elaborate prayer to the same effect:

> In the future that we may continue to hold each other with the turquoise hand,
> Now that you may return to the place from which you came,
> In the future as time goes on that I may rely on you for food
> To the home of the dawn you are starting to return.
> With the jet hoofs you are starting to return.
> . . .
> In beauty you have arrived home.
> In beauty may you and I both continue to live.
> From this day you may lead the other game along the trails that I may hunt,
> Because I have obeyed all the restrictions laid down by your god in hunting and skinning you;
> Therefore I ask for this luck that I may continue to have good luck in hunting you.[16]

There was a difference of opinion on whether one should breathe the breath of a dying animal; Pueblo Indians did this to partake of "the mysteries of life," but Navajos strictly avoided it, as "one would be sick at once."[17] After the kill, the pelt was cleaned of blood and dirt. A Tsimshian put his knife at the bear's chest and sang "the bear's mourning-song."[18] Then the hide might be removed, laid back on the animal in reverse direction, head to tail, with an appropriate prayer. Since they often killed hundreds of animals at once, the Plains Indians did not always apologize to each individual one, though they did so to the herd or guardian spirit. But in the case of species that were caught one at a time, like the eagle whose feathers were desired for ceremonial decorations, the Plains hunter would ask for pardon. In any of these cases, if the proper ritual were neglected, the animal's spirit would retaliate in any of several ways: by putting its skin back on and escaping, by making the hunter fall ill or have an accident, by stealing away his hunting power, by refusing ever to return and be hunted again, or by warning others of the same species to desert his hunting territory.

The dead creature's body, and all its parts, had to be treated carefully for the same reasons. It might be tied and carried with special bindings or a beautifully decorated pack strap to please the being and prevent harm to the hunter. Great care was taken with a bear's eyes; they were returned to the den if possible. When the dead animals were brought home, they were treated with further respectful ceremony; the head of each animal was again turned in the direction thought proper by the particular tribe. Words of welcome were said, and offerings made of corn, cornmeal, and feathers and other beautiful things. This was done for animals as small as rabbits. The first one of a certain species to be killed in a given hunting season might be treated with special honors. Here again, bears claimed the most elaborate rituals. When a bear's carcass arrived in the village, it was welcomed with songs and dances. When a bear was brought into a Micmac wigwam, it came in through a special door that was used only for that purpose. The Nootka addressed the bear's body as "chief" and placed it in a chief's seat with a chief's hat on its head. After killing a bear, a Dakota chief named Wawatam held a ceremony for his people in which he grieved that it had been necessary for him to kill a "friend." The hunter himself usually went through a rite of purification, for dead animals have power.

Practically no waste was allowed. Some parts of animals and birds were kept for use in ceremonies. The meat of wild animals was boiled carefully; it was believed to be dangerous to burn it, or eat too much. The bones were never given to dogs to chew, and were kept away from streams, fire, and menstruating women. Often they were placed in a shrine. The skin and head received particular attention; the skull might be fixed high in an evergreen tree to keep it from dogs and pollution of various kinds.

CHAPTER III

The hunting of the great whales by the Nootka and Makah offers an illustration of the important part played by ritual in the whole process. Most whales were longer and much larger than the hunting canoes, and were regarded as beings of great power, particularly the killer whales, which were rarely hunted. The Haida word for "power" is *sgana* or *sganagua;* for "shaman," *sgaga;* and for "killer whale," *sgana;* the concepts as well as the words were obviously related. The Nootka chief, Maquinna, would spend a whole day in self-discipline, song, and prayers intended to call, calm, and placate the whale. When the leader of the hunt went forth with his crews, while his wife lay still at home, he addressed the whale as "noble lady," and urged her to cooperate by moving her vast and terrifying body in ways that would help the hunters.[19] After a successful kill, he would be allowed to keep the whale's eyes and the best strip of blubber, which he would place on a rack like a shrine and decorate with eagle feathers to mollify the whale's spirit. The people of the village gave the whale's body a festival of welcome, offering it fresh water, eagle's down, and other symbolic gifts. A similar welcome awaited the whale that had beached itself. Some men were believed to have the gift of being able to call whales and cause them to do this.

As we have seen, each "tribe" or species of animals was considered to have its own Chief, Master or Keeper.[20] Some were more powerful than others, but all demanded that their children be treated with care and respect. There were, for example, many raccoons, but there was also The Raccoon, the one who cared for all raccoons and epitomized the power possessed by raccoons. Sometimes the Keeper was seen as a particularly large and beautiful member of the same species, perhaps with special markings. Often he had human characteristics as well. Or he might be another, associated animal. In at least one tribe, the moosefly was called the Overlord of Fish, and his stings were intended to warn people against wasting the fish he protected.[21] When a deer was killed, "somewhere in the mountains, there was a greater deer spirit whose power had been damaged to some extent, and an element of disorder thus introduced into the world."[22] The result would be suffering for the hunter unless the proper ritual amends were made to the deer spirit. The Cherokees said that Little Deer, the being who is Master of all deer, always came to ask the spirit of a slain deer if the hunter had apologized properly.[23] If not, he would strike the hunter with rheumatism.

The Powerful Animals

In the Southwest, the deities associated with wild animals were well differentiated: some, usually female, were primarily protectors of animals, and others, usually male, were patrons of the hunt. A Mother of the Beasts appeared widely, watching over her animal children and encouraging their reproduction. If approached in the right spirit and with proper ceremonies, she might willingly give some of them to human beings who needed them. The male Masters of Game taught skills and hunting ceremonies, and helped hunters find their prey. Such a one was the Navajo Talking God, who observed hunters to see if they obeyed the rules of the hunt and respected the animals. If they did, he would call the game to them. His companions were birds: "When we see Bluebirds or other small birds, we know he is near."[24] One story tells of Deer Owner, a perverse Master of Game who kept all the animals for his own use until he was bested by a hero, forced to release them, and teach his skills.

Animal spirits were believed to come among the people to participate in dances. The Buffalo Dance, Deer Dance, and Eagle Dance with their striking imitations of animal behavior are well known; there were many more. The Pueblo Kachinas often represented animals and were believed actually to embody their spirits. The feeling of comradeship for animals combined with respect and awe of their power to explain the use of animals in ceremonies such as the live snakes that are carried in the Hopi Snake-Antelope Ceremony. Having been well treated by the people, they would return to the natural environment with the petitions of the people for rain. Prayers and offerings were made for the fertility of all wild animals. The animal gods who were the most skilled predators also served as patrons of the hunt, and were especially respected. If the hunter practiced the ritual carefully, he could expect help both from the game animals, who would present themselves to him, and from the hunting spirits like the wolf, who would round up other animals and drive them toward him. At Zuñi and elsewhere, these helpers were the guardians of the six cardinal directions: north, Mountain Lion; west, Bear or Coyote; south, Badger or Bobcat; east, Wolf (the great Hunt Chief); above, Eagle; below, Mole or Shrew. The assistance of Wolf, Bear, and Mountain Lion was especially sought by hunters. Wolves themselves were not usually hunted; according to the Coast Salish, if a wolf were shot, the hunter would be surrounded by other wolves and killed, unless he gave them an acceptable apology. There was also a Great Keeper, owner of all the animals, a being sometimes seen as half-human, half-animal, particularly half-bear

CHAPTER III

(stories of humans marrying animals and having children, without necessarily realizing who their partners were, occur in many tribes). Northeastern Indians would tell of glimpsing the Great White Bear, the Master of Animals, in the forest as a terrible omen. He took care of the animals, would warn them to leave if they were not treated with proper respect, and could cause a hunter to die by "accident." With this perception of animals, Indians would not kill them except in genuine cases of need, and then only with the greatest caution.

The result of this attitude was that the Indian hunter was not a "miner" of the game resource, but a conservator. All Indians who hunted and gathered followed the twin principles of wise use: they took only what they needed, and they completely used everything they took. The hunting ideal of "use all you take" was practiced everywhere in aboriginal America. Indians felt strongly that no part of any creature's body should be wasted, and a use should be found for everything. In the case of a moose, for example, the hide could be used for tent coverings, moccasins, robes and leggings; the meat could be eaten or preserved by smoking or drying, and the entrails and fat consumed as prized delicacies; the bones could be cracked for marrow, rendered for grease, or fashioned into implements; the antlers, teeth and the bones could become tools, ornaments, or counters for games. As little as possible was thrown out.

The principle of "do not take all" was embodied in many Indian practices or hunting rules that were learned as sacred teachings. The examples are many. The Yuroks killed buck deer only, allowing a doe to be taken only in emergency. In some tribes pregnant animals were spared, as were all females during breeding season. The ability of Indian hunters to be very selective of the animals they took is amazing. Many were runners good enough to pursue their prey on foot until it slowed from exhaustion, and often smothered it with their hands, to keep from damaging the hide with an arrow mark.

Often the young animals were spared so that they could grow up and be hunted as adults. The Pomo in California's Clear Lake country would not take ducklings until they were at least half-grown.[25]

Certain species were never killed, although just which ones varied from tribe to tribe. For example, the Utes always spared the gray jay, who raucous calls protect small animals and help hunters find the predators. Predators were hunted, but never persecuted with the idea of destroying them all, as some White hunters later tried to do with disastrous results. Meat which had been slain by a predatory animal was never eaten. The usual reason for not killing a species was its possession of special power. In pre-contact times Tlingits would not kill land otters or use their fur, because they were shaman's guardian species. It was widely felt that bears, eagles, snakes, and coyotes should be hunted only for unusual reasons. The killer of a bear or eagle might be treated in the same way as the killer of a human enemy, with long ceremonial purification and initiation into warrior societies. "It was just like killing a man to kill a bear,"[26] said a Navajo, and an Isleta man remarked, "Bear is a person."[27] Still, a bear might be killed if it were a habitual raider of cornfields, or if people were starving. In such cases, the prescribed rituals and apologies were most stringent. The Delaware Bear Ceremony included the sacrifice of a bear, who had indicated his willingness in a dream, at the door of the longhouse. Twelve men noted for their good behavior in hunting were chosen to bring in the bear. After the sacrifice, the bearskin (along with the skull, an object of special significance among many tribes) was placed around the center post of the longhouse as the climax of a processional ceremony including sacred songs and dances. The Dark Dance of the Iroquois, an autumn hunting feast which the animal spirits were invited to share with the people, was marked by the ceremonial eating of bear's flesh. Here the bear hunt was held only for ceremonial purposes.

Clans were usually named after animals, birds, or plants, and the creature so honored was treated with special reverence, both in ceremonies and prohibitions against hunting, or eating at least certain parts of the totem animal, which they regarded as their kinfolk. Members of the bear clan, for example, might refuse to eat bear meat or otherwise harm their totem. Bear designs would decorate their implements and clothing. Clan duties were sometimes rationalized in keeping with the qualities of the clan animal. The Bear Clan might provide chiefs, for example, because bears are powerful and yet peaceful, as Badger Clan members could be chosen to impersonate the Badger in ceremonials.

Other "game laws" limited the number of animals that could be taken at one time. When Hopi hunters caught a herd of mountain sheep, they always released two, a male and female, "to make more."[28] The Zuñis made a point of allowing the leader of deer herd to escape, since he had brought his children to them. Game surrounds always allowed the first few animals to get away as the circle contracted. "The hunt was fruitful, but it was no mass slaughter," said an eyewitness.[29] The Tlingits would not allow a marmot hunter to set more than eight

CHAPTER III

deadfall traps at one time. Among the Plains Indians, elks' teeth were much admired as decorations on woman's dresses. But since each elk provided only two of the prized teeth, imitations were made from bone or shells.

Other customs limited the hunting season for different animals. Certain months, as at breeding season, were avoided. Hopis said that rabbits wept if hunted during the December moon, so they left them alone then. Tsimshians said that porcupines "must not be smoked out of their dens in winter."[30] For agricultural tribes, hunting was limited to times when the crops did not need attention. Zuñis would not hunt deer longer than four days in a row, even if they were having great success, four being a sacred number of completion.[31] They hunted turtles at the sacred lake only once every four years. The Gosiute of Nevada had the custom of waiting twelve years between antelope drives, to allow the herd to reestablish themselves.[32]

The "single prey species" principle was common among many tribes, that is, that a good hunt should be only for one chosen species of animals, and should only use the one way of hunting selected for that hunt; to take other opportunities of killing that presented themselves might bring serious consequences. This made a hunt into specific ecological tool, not an indiscriminate bagging of game. The Yuroks, for example, prohibited the taking of quail and grouse in the season when deer were hunted.

The way a species was protected by the customs observed when hunting it can be seen in the case of eagles. Because their feathers were indispensible in ceremonies, eagles were highly desired, but hunting them was dangerous. Some hunters hid themselves in camouflaged pits, with a rabbit as bait, and grasped the eagle's feet when it swooped down for the prey. The first eagle to be caught might later be released. Some Navajos plucked their captive eagles but eventually released them after the feathers regrew. Hopis took eaglets from nests on high cliffs and left valuable offerings in their place. Each nest was known and owned by a particular clan. No eggs were ever taken, and only two eaglets were removed from a nest. Once back in the pueblo, the eagles were adopted into the tribe, fed daily by hand, and held captive by leg thongs until the day when they were sacrificed. The plucked bodies received careful burial in a tribal cemetery, and prayers were said that their spirits might fly back to the cliffs and be reborn. Among the Papagos, "it was not thought right to kill more than one eagle in one year," certainly a form of conservation.[33]

The care with which Indians treated wild animals resulted from the fact that they depended on the ecosystems within which they lived. Hunting was never a sport. It was the necessary means of feeding the tribe, and the tribe would always depend on the animals, birds and fish that lived in its own land. The usefulness of conservation practices, indeed, their virtual necessity, is evident. Indians held to the principle that hunters should never kill more than was needed. Whites who observed Indians in early times very rarely reported any instances of wasteful hunting, and then only as remarkable exceptions. Indians cared for the living creatures they hunted, and protected them from overexploitation. The great student of northeastern forest Indians, Dr. Frank G. Speck, reported some remarks of Aleck Paul (Osheshewakwasinowinini), Second Chief of the Temagami Band of Ojibwas, that illustrates this attitude:

> We Indian families used to hunt in a certain section for beaver. We would only kill the small beaver and leave the old ones to keep breeding. Then when they got too old, they too would be killed, just as a farmer kills his pigs, preserving the stock for his supply of young. The beaver was the Indian's pork, the moose his beef, the partridge his chicken; and there was the caribou or red deer — that was his sheep. All these formed the stock on his family hunting ground.[34]

Here an Indian finds a good comparison between the attitudes and practices of the Indian hunter and not the White hunter, but the White farmer and his treatment of domestic animals.

Aleck Paul's remarks also indicate the existence of a widespread institution among the Indians that operated to maintain conservation: the hunting territory. Each family or band had a recognized area of land as its own hunting ground. Within that territory, they tried to conserve the supply of animals, birds and fish. They took what amounts to a continuous game census, always knowing how many there were of each kind, and where they were located. Usually the territory was divided into sections, perhaps four in number, and the Indians would hunt in only one section each year, giving the game in the other sections a chance to recover. They might also reserve a fifth, central section as a reserve where they would never hunt at all, except in time of great need.[35] This is a brief description of the "family hunting ground" system used by the Northeast forest tribes, the Utes of the Rocky Mountains, and elsewhere.

CHAPTER III

Tribes that did not use this system achieved the same result through tribal decisions. The Choctaw tribe in the Southeastern forest made an annual determination as to how much game could be taken by each family, and where.[36] They kept track carefully, and no one was allowed to kill too many. Each tribe defended its hunting grounds against excessive use by others. Indians were hospitable and would allow reasonable hunting and fishing by guests who came in friendship, but indiscriminate killing could be given as an acceptable reason for war. These hunting practices must be understood within the context of the Indians' respect and esteem for animals. No creature should be needlessly destroyed or wantonly exploited. The Apaches felt bad if they even stepped on a spider. For people with such attitudes, hunting was a ritual, and cruelty to animals or mistreatment of their remains was strictly forbidden to hunters.

This being the case, why do so many non-Indians today have the idea that the Plains Indians were wasteful in their buffalo hunts? This common misconception needs to be considered in detail.

In the days before the horse, Indians had to hunt buffalo on foot, surrounding them, causing them to jump over high bluffs or go into carefully built "pounds." In winter, they could be found mired in deep snowdrifts. Rarely were stalking and fire drives used. The animals were killed with bows and arrows or spears. The introduction of horses meant that "running" the buffalo became the usual hunting pattern, with riders able to select individual animals and pursue them. Bows and arrow remained the preferred weapons, since reloading a single-shot rifle after discharging it on horseback was very time-consuming. In the same time, several arrows could be loosed at the buffalo.

It has too long been commonly said that Indians were wasteful in their buffalo hunts, or that they were in the process of hunting the herds to extinction. It is true that the thought that buffalo might disappear from the earth was inconceivable to Indians before the fact began to be apparent in the late nineteenth century. Many of them held that the buffalo were renewed from under the earth each spring, or, like the Forest Indians' belief concerning the deer, that buffalo bones would become living animals again and allow themselves once more to be hunted. But Indians were too familiar with times of famine in the absence of the herds from their own area to believe that buffalo were inexhaustibly available to them. Conservation was a wise method of tribal survival. Hunted in the numbers required for Indian needs, the

The Powerful Animals

buffalo were indeed inexhaustible, but not always available. It was the White commercial hunters, not the Indians, who decimated them.

During the years when buffalo were most numerous, the Plains Indians protected them carefully, for in so doing they were safeguarding their own food supply. Frank Gilbert Roe, the definitive historian of the North American buffalo, has collected the comments of early White observers on this subject, and many of them suuport the statements made above. In 1661, Pierre Esprit Radisson observed, "This place hath a great store of (buffalo) Cows . . . The wild men (Indians) kill not except for necessary use." Hennepin noted in 1689, "These (buffalo) Bulls being very convenient for the Subsistence of the Savages (Indians), they take care not to Scare them from their Country; and they pursue only those whom they have wounded with their arrows." Westerners of the mid-1880s were prone to exaggerate the Indian's sins in order to justify White conduct, which makes a comment like R. B. Marcy's, written in 1850, all the more valuable:

> (The buffalos') only enemy then was the Indian, who supplies himself with food and clothing from the immense herds around his door; but would have looked upon it as sacrilege to destroy more than barely sufficient to supply the wants of his family.[37]

And the Sioux City Journal of May, 1881, added: "To the credit of the Indians it can be said that they killed no more than they could save the meat from."

No tribe permitted unauthorized buffalo hunting, and this rule was enforced by military-police societies like the Cheyenne Dog Soldiers. Any men who went out by themselves to kill buffalo outside of the announced tribal hunts faced severe punishments: Depending on the seriousness of the violation, they could receive harsh reproofs, be beaten even to suffering broken bones, have the meat taken away, their tepees burned, their horses shot, or see all their possessions confiscated. Such a list of punishments shows that the rules were sometimes broken, but not often. When the Assiniboins acquired rifles around 1800, they established a rule forbidding their use to kill buffalo.[38] Similar rules against hunting other animals or gathering wild plants before the proper season were also enforced by the "soldier" societies.

Indian hunting methods were not such as to wipe out great herds of buffalo. In early surrounds, some usually managed to escape. A 1691 account says that the Indians surrounded a herd, "keeping the Beasts

CHAPTER III

still in the middle and so shooting them till they break out at some place or other and so get away from them."[39] Corralling often failed because no buffalo could be induced to enter. Many Tail Feathers, a Piegan, was reportedly thanked in a vision by the buffalo and given the right to wear a special red war bonnet because he had destroyed a corral and thus saved some of them.[40] Hunting on horseback proved to be efficient, since a buffalo could be killed quickly with one or two arrows, but an average kill of one or two buffalo per hunter was usual for a hunt. Those with the best horses might manage to kill four of five apiece, but others more poorly mounted might not manage to kill any. The numbers killed in hunts actually observed were not spectacular; Buffalo Jones watched a Pawnee hunt in 1872 where only 41 animals were killed out of a herd of 2,000 in "a day's work."[41] Weasel Tail's account of an 1874 hunt said only 33 were driven into a pound by horseriding Piegans, and killed.[42] The corral was not left in place. but afterwards burned as firewood. Once in a pound, all the captured buffalo were usually killed, since any surviving animals would present a danger to the hunters, and it was believed that if they were released, they might warn others to stay away. Hunting was sometimes used as tool in herd improvement; diseases of buffalo were not common, but during an epidemic of mange, every infected animal found was killed. Indians themselves reported in the mid-nineteenth century that they observed no general reduction in the number of game animals until after the period of White contact.

But what about the museum exhibits and textbook illustrations that show Indians driving whole herds of buffalo over cliffs, based on the evidence of huge piles of bones found in many widely separated places all over the Plains? At first glance, these seem to indicate appalling waste, but on closer examination, this impression is shown to be false. Most "buffalo jumps" date from before the introduction of the horse and the more selective hunting that accompanied it. But even ancient mass kill sites of bison show signs that the animals were carefully butchered, and every useful part removed and carried away. Joe Ben Wheat of the University of Colorado Museum excavated a bison hunt site in the Colorado Plains where, in about 6500 B.C., almost 200 buffalos were driven down into a steep ravine and killed.[43] The bones of most of the animals were found in "butchering units," often in piles consisting of the same parts from many animals, indicating systematic handling. The tail vertebrae were missing, showing that the hides had been removed, probably filled with meat destined for preservation by drying, and possibly dragged by the tails. Of the 193 skeletons found in the excavation, 75% were fairly completely dismembered, and another 18% partially so. Only 7% might be called wasted, and these were mostly in parts of the site where accessibility would have been difficult. This does not present a picture of wasteful slaughter but in fact, quite the opposite.

In describing a similar but older site (about 8000 B.C.), near Casper, Wyoming, where buffalos were trapped among sand dunes by hunters, George C. Frison of the University of Wyoming says, "There is . . . a strong indication of effective utilization of the meat products in communal kills."[44] He points out that the buffalos were completely butchered, many bones stacked, and even about half of the long bones were removed, presumably for their marrow content. This pattern of careful butchering continued; Cabeza de Vaca in the 1530's described the Indians' complete use of buffalo carcasses.

Some comments by early White travelers indicated that Indians did not always utilize their kills completely, sometimes taking only the tongues and the best pieces, but this seems to have occurred only in exceptional circumstances, and many Whites were anxious to portray Indians as wasteful, lazy and improvident. Usually many bones were left behind at the kill site, since they were too heavy to be carried easily, and later passers-by, encountering heaps of bones and offal, received an impression of Indian wastefulness.

The evidence of labor at the kill sites, and complete use of the products of the buffalo, leads to other conclusions. Anthropologists have listed close to a hundred major Indian uses for different parts of the buffalo, everything from horn spoons and hide boxes or chests to buffalo hair weaving and water buckets made of the paunch.

Waste of buffalo meat was minimized by its preservability. Great quantities were dried as jerky or combined with fat, marrow, and the paste of crushed wild cherries to make pemmican. One man could butcher from five to twelve animals in a day, and two horses were required to carry the meat removed from a buffalo. As Edwin Jones said in 1820, "Every eatable part of the animal is carried to the camp and preserved, except the feet and the head."[45] Even those parts were sometimes preserved for ceremonial use. Something that one might think of as waste, stomach contents, were actually vegetable food in a digestible form, rich in vitamins, and they were often consumed.

Another factor preventing waste and assuring full use was the Plains Indian ethics of sharing. Generosity was an honored ideal, especially for chiefs, and the presentations of gifts, including gifts of food, was an institution. Provision was made for those in need as long as there was anything to give.

Recognized tribal hunting territories, and their defense against intruders, served as a mechanism to conserve the buffalo herds. Plains warfare was almost as stylized as a medieval tournament, and was seen by its participants not so much as a way to kill enemies as a means of demonstrating personal skill and bravery, but it also represented competition for good hunting grounds. This factor became more important as White pressure in the east, and the mobility provided by the horse, brought tribes formerly on the periphery out into the Great Plains. Plains Indians lacked a concept of land as disposable private property, but an area of land, with the animals on it, belonged to the tribe. They usually spoke of "our buffalos" in the same protective way in which a rancher might later speak of far-flung unfenced herds as "my cattle."

The Indians were living in ecological balance with the herds of buffalo. Before the White onslaught upon their ecosystem, they were not involved in a process of wasteful slaughter leading toward extinction, but were practicing conservation according to their own time-honored principles: to repeat, "take only what you need and use all you take." The vast herds which flourished on the Great Plains in undiminished abundance as long as the Indians were their caretakers, constitute the best evidence in support of this assertion.

As we noted above, animals that had been angered by human actions were believed to be able to send sickness. This was true whether the actions were deliberate or not, and usually they were inadvertent; hardly any Indian would have done such a thing on purpose. A Pueblo Indian shot a bear by mistake; a Navajo woman camped on a deer trail; an Apache drank water a bear had bathed in; each became "sick from the power of the bear" or deer.[46] If a man negligently mistreated so much as an anthill, the ant spirit could send ants within his body to torment him. A Papago, might have mistakenly smashed a rodent's burrow, wasted meat, thrown bones away or let them come into contact with a menstruating woman, or killed an animal with unnecessary cruelty, and suffered the consequences. Almost any creature could send its own particular kind of illness: a cough, for example, came from deer. For every animal, there was believed to be a plant that could be used as medicine against the animal's disease.

Even more often used was the help of a medicine man who had gained access through a vision to an animal spirit helper; his special songs and talismans had power over disease and other natural phenomena. Among the Navajo, the medicine man was a specialist and a carefully trained priest who spent years learning chants, sand paintings and ceremonies that took several days to execute correctly. He would begin treatment only after a diagnostician had decided which chants would be effective in the particular case. Pueblo Indians depended less on visions and more on the careful performance of ceremonials by healing societies (which might also be war or hunting societies) with beast gods as their patrons. The requisite animal was often impersonated in these ceremonies. The member "became one person" with the animal. The Hopis tended to regard mistakes in ritual, and breaking of ceremonial taboos, as the primary cause of disease.

The principle that animals could not be mistreated without serious consequences prevented needless cruelty and thoughtless destruction. They did not allow children to play with wild animals such as mountain goat kids or trout, put frogs into fire, or throw rocks at birds or squirrels. If a Kwakiutl child were to kill even a squirrel, the father would say a prayer expressing his apology for the "accident," and asking for some of the squirrel's quality of energy. To kill an owl would be worst of all, said the Kwakiutl, because an owl is the "mask" that waits to fly away as the outer garment of the soul of a human being, and the result would be the death of that person. An owl might be caught and given offerings of ochre, but would always be released unharmed.

Speaking of the treatment of animals reminds us of the domestic animals that Indians kept in their villages and camps. Only one, the dog, was known in aboriginal times. It was a beast of burden, not a pampered pet, but was cared for. The Assiniboins, for example, carried water for their dogs in buffalo paunches on treks through dry country. The Northwest Coast Indians sometimes used them to hunt with.

The horse, which had become known throughout the Great Plains in the course of the eighteenth century, fulfilled a unique role in the Plains Indian world view. As Frank Gilbert Roe so aptly put it, "The horse widened the Indians' spiritual horizon."[47] While recognized as a domestic animal, it was also seen as a creature of great power, and thus regarded with the respect and religious awe accorded to wild animals. This ambivalence is contained within the Sioux expression for "horse," *shonka wakan*, the "power dog," "spirit dog," or "mystery dog," as it has been variously rendered in English. In some other languages, "horse" is expressed as "elk-dog." Horse medicine, received through

CHAPTER III

visions of horses, was regarded as most powerful. Wild horses were captured by medicine men who had received the power to attract them. Anyone who has read *Black Elk Speaks*[48] has been impressed by the dominant role of horses in a Sioux's personal revelation. A boy of the Hidatsa tribe expressed both sides of the Plains Indians' paradoxical attitude toward horses by saying to some that he happened to be watching, "You are my gods. I take good care of you."[49]

Plains Indians treated their horses well. As Larocque noted of the Crow tribe in 1805, "They are very fond of their horses and take good care of them; as soon as a horse has a sore back he is not used until he is healed."[50] A practical kind of veterinary medicine was used to treat horse ailments. Cottonwood bark and grass were gathered and fed to horses, and the best steeds were often kept inside the lodges on cold winter nights. In 1852, Sanaco, a Comanche chief, said of his horse, " I love him very much."[51] They ran their horses hard in battle and the hunt, and sometimes cut their nostrils to give them better wind, but these things were done under the pressure of necessity; Wolf-Chief, a Hidatsa born about 1849, remarked that he never whipped his horse except when pursued.[52] The Indian received the horse's gift of speed with gratitude and admiration.

Other domestic animals were readily adopted by the Southwestern Indians, particularly sheep and goats. The eventual impact on the land would be catastrophic. Fields had to be fenced to keep them out, but fences and tradition conservation measures could not counteract the long-term effects of erosion caused by over-grazing. Indians, especially the Navajos, came to have great affection for their domestic animals, treating them almost like family members, and using the word "love" to describe their feeling for them.[53] Their attitude toward sheep and goats was not an extension of the respect and awe of power they felt concerning wild animals, but more like the treatment accorded the dog. Legend assigned them a very early creation, but not necessarily the same origin as that of wild animals and human beings. They could be marked with signs of human ownership. Indian handling of pets sometimes appeared callous in White eyes, but there were strong feelings against the mistreatment of horses, dogs, and other domestic animals.

Fishing by Indians reflects the same feelings as hunting and many of the same rules. For example, the Chumash had uses for every part of shellfish like the abalone, casting aside as little as possible. Of all the North American cultural areas, the one where fishing was most important was the Northwest. There, the greatest single gift bestowed by the natural environment, and the mainstay of the Indians' life, was the yearly return of the several species of salmon to the streams to spawn and die. More than anything else, salmon represented the abundance of nature for the Northwest Coast Indians. They took them with nets, harpoons, spears, hook and line, built weirs across the streams, and constucted clever traps of poles and withes. They took salmon in numbers great enough to smoke and preserve large stores of them for the rest of the year, freeing themselves from constant concern about hunger. They refrained from taking more than they needed for one year; most believed, like the Puyallup, that the salmon would take away the soul of a wasteful person.[54] People that built weirs across the streams kept them closed only until they had caught enough; the bulk of the salmon were allowed to run through. This was done in part to allow upstream villages their share; conservation rules of this kind have survived in some places into recent times. Water pollution was prevented by ritual prohibitions against putting such things as urine or blue hellebore (a plant poisonous to fish and sea mammals: they did not fish with poison) into the sea. During the time before the White occupation, the depletion of salmon runs was never a problem in the Northwest.

The salmon were believed to be a numerous host of spirit beings who had their village out under the western ocean. Those Indians who visited them in visions saw them living in great houses like human beings. Their annual pilgrimages to the mountain streams was seen as a voluntary sacrifice for the benefit of their human friends. Though they seemed to die, the spirits had simply removed their outer "salmon robes,' and journeyed back to their undersea homes. But they would return again only if their gifts of flesh were treated with careful respect. When caught, they might be clubbed once, but not more. Children were taught not to play with salmon; serious cautionary tales warned of the dire consequences provoked by a boy who had poked out a salmon's eyes with a stick.[55] The meat could be broken by hand or cut only with the most traditional knives, such as those made of shell, certainly not with metal. The bones and other offal had to be kept from women who were menstruating, and from dogs or birds; they were put back into the water so that the salmon spirits could find them again. Those who were believed most able to affect the salmon runs: twins and their parents, girls at puberty, and mourners, were subject to

CHAPTER III

ritual restrictions for various periods of time, during which they were isolated, allowed to eat only certain foods, and forbidden to hunt or fish.

The First Salmon Ceremony was observed everywhere along the Northwest coast. For the Yurok and others, it symbolized the renewal of the world's creation. Many groups also held ceremonies for the first fish taken of other species, the first deer, first berries, or even the first acorn in the southern reaches, the first salmon (or the first four salmon) received the most elaborate rites, though they varied from place to place. Usually the salmon were laid with their heads pointing upstream, on a newly woven mat or a cedar board, often under a special shelter, and sprinkled with the down feathers of birds. A formal speech or prayer of welcome was intoned, as in this Kwakiutl example:

> Oh friends! thank you that we meet alive. We have lived until this time when you came this year. Now we pray you, Supernatural-Ones, to protect us from danger, that nothing evil may happen to us when we eat you, Supernatural-Ones! For that is the reason why you come here, that we may catch you for food. We know that only your bodies are dead here, but your souls come to watch over us when we are going to eat what you have given us to eat now.[56]

On behalf of the fish, the people responded, "Indeed!" The salmon were offered fresh water symbolically, after their long journey through the salt sea. The first salmon were then cooked and divided in small pieces among all the people present as a kind of communion. The celebration, often several days in length, included feasting, gift-giving, torchbearing processions, dancing, and singing. During all the long ceremonial of welcome, countless salmon were allowed to pass upstream to the spawning grounds, and thus the ritual actually helped to assure the continuation of the salmon runs.

Fish of many other species were valued, especially the eulachon (candlefish), whose bodies were extremely rich in an inflammable and edible oil. There were prayers, ceremonies, and attendant ritual prohibitions for them too. As the Tsimshian netted eulachon, they shouted to them, "You are all chiefs." The first one was placed on an elderberry frame, addressed with special words, and its oil was pressed out against the women's bare breasts, as any other method would "shame" the eulachon.

The Powerful Animals

The appearance of surf smelts on the beaches, fish that can be caught by hand, is predicted today by those who know how to consult tables of the tides and the moon's phases, but some Makahs were provided with similar predictions by their guardian spirits. Their sense of the supernatural power wielded by fish, as by other animals, led the Indians to treat them with great consideration. When a Bella Coola, for example, caught a bullhead in a eulachon net, he released it, saying, "I have saved you; please do the same for me."[57]

A final and very important aspect of hunting that supported the conservation practices of the Indians was the ethics of sharing. A hunter killed not for himself alone, but for his clan and tribe, with special care that a portion should go to the old and needy. Indian societies showed an absence of greed; their traditions taught generosity with food and possessions, so that when one person had food, all had food, and when one starved, all starved. Thus the hunters would harvest only what the community needed, avoiding the depletion of game that might mean hard times for everyone.

Along the Northwest Coast, the ethics of sharing, so characteristic of American Indians in general, achieved an extreme expression reflecting the abundance of the natural environment. One of their most notable social customs was an elaborate winter gift-giving ceremony called the potlatch, in which a man of standing showed his position and wealth by distributing large quantities of food and almost every form of movable property. Among the things given away, or sometimes even ostentatiously destroyed, were animal pelts and wool, robes, strings of dentalium-shell money, large copper plates often with elaborate decorations worked into the surface, wooden bowls and ladles, cedar house planks, and, in later times, blankets. The recipients were members of other families in the village, or the people of neighboring groups. Potlatches were considered necessary for advances in status, the assumption of right to houses and resource territories, and the celebration of important events in the lives of children and adults. They seem to have been very common, and the amounts of food and property redistributed through them were truly impressive. The rivalry potlatch, in which two claimants to the same honor each tried to outdo the other in generosity, and often impoverished themselves temporarily by giving away or destroying all they possessed, seems to have occurred only rarely. The potlatch can probably be understood best in terms of ecological relationships as a way of equalizing the food supply and other resources between families within a village, and between

CHAPTER III

villages.[58] Even in an area of abundance, not every item of food could have been available in surplus in every local area in every year, so potlatching was a way of redistributing the surpluses which did occur. A family or village that gave at one time could expect to receive at another.

In other parts of North America, sharing was simpler, but everywhere it was practiced and approved. As the Dakota Sioux elder, Ohiyesa (Charles Eastman), stated this central principle of Indian ethics,

> The native American has been generally despised by his white conquerors for his poverty and simplicity. They forget, perhaps, that his religion forbade the accumulation of wealth and the enjoyment of luxury. To him, as to other singleminded men in every age and race, from Diogenes to the brothers of Saint Francis, from the Montanists to the Shakers, the love of possessions has appeared a snare, and the burdens of a complex society a source of needless peril and temptation. Furthermore, it was the rule of his life to share the fruits of his skill and success with his less fortunate brothers. Thus he kept his spirit free from the clog of pride, cupidity, or envy, and carried out as he believed, the divine decree — a matter profoundly important to him.[59]

To most people who have tried to imagine how human beings can live in harmony with the natural world, a simple life and a sense of sharing the earth not only with other people but also with all living creatures seems to be an essential part of the attitude that is required. Indians had such a sense of sharing with other members of the tribe and with the powerful animals.

The Powerful Animals. A Ute Indian stands beneath an eagle that is held captive in a tree. All tribes venerated the eagle as a spirit being of great power.
　　　　COURTESY SMITHSONIAN INSTITUTION NATIONAL ANTHROPOLOGICAL ARCHIVES.

The Powerful Animals. Hopi Indians at Walpi are shown dancing with snakes, which are asked to return to the natural environment and ask the powerful spirits of nature to fulfill the needs of the people which, in this case, is rain.

COURTESY MUSEUM OF THE AMERICAN INDIAN, HEYE FOUNDATION.

The Powerful Animals. The Karok Indians of northern California perform a Deer Dance, in which the white deerskins are moved in a manner that is evocative of live deer. Photo circa 1908.

COURTESY MUSEUM OF THE AMERICAN INDIAN, HEYE FOUNDATION.

The Powerful Animals. Some captive, immature eagles and small tihu *offerings were photographed in Walpi, a Hopi pueblo, in 1921, by Emery Kopta. The Hopis collect eagle chicks from nests and raise them for ceremonial purposes.*

COURTESY MUSEUM OF THE AMERICAN INDIAN, HEYE FOUNDATION.

The Powerful Animals. The bear, universally recognized by Indians as a source of immense spiritual power, is represented on totem poles in the Haida Indian village of Massett, Queen Charlotte Island, British Columbia. The grizzly bear carving in the foreground is a totem of Chief Edensaw. Photographer F. Maynard, circa 1880-1885.

COURTESY SMITHSONIAN INSTITUTION, NATIONAL ANTHROPOLOGICAL ARCHIVES.

IV

The Plant People

IF ANIMALS were spirit people in the Indian world, so were plants. Tatanga Mani, or Walking Buffalo, a Canadian Indian of the Stoney tribe, explained it this way:

> Did you know that trees talk? Well they do. They talk to each other, and they'll talk to you if you listen. Trouble is, white people don't listen. They never learned to listen to the Indians so I don't suppose they'll listen to other voices in nature. But I have learned a lot from trees: sometimes about the weather, sometimes about animals, sometimes about the Great Spirit.[1]

Plants were beings that had agreed to be of special help to mankind, and must therefore be treated with consideration and respect. The tribes that hunted also gathered wild plants. There was a division of labor by sex in most tribes that was perhaps not very strict. Men usually hunted; women usually gathered. This did not necessarily mean that the men's task was thought to be more important. Indeed, some scholars point out that the actual weight of food collected by the women was probably greater than that killed by the men. And exceptions occurred: women sometimes hunted and men often gathered plants, particularly if they were medicine men searching for powerful herbs to use in ceremonies. But the gatherer of plants for food and medicine was usually a woman; she had the same attitude toward plants that the hunter had toward the animals, and her actions were analogous. Plants had to be gathered in just the right way.

Since Indians knew that plants were alive in the same way that animals and human beings are alive, they had no reason to be vegetarians. Eating plant foods involved some necessary killing, just as eating animal foods did. Wild plants served the Indians as food and as medicines. Berries, roots, bulbs, shoots, inner bark, nuts, and acorns

CHAPTER IV

were used where they were available. The wise elders whose job it was to know such things could tell the uses of every plant in the ecosystem. The Cheyennes, for example, were known to use about forty different fruits, roots, stalks, and buds for food, and the number of plants that had medicinal value was in the hundreds. Such intimate knowledge of wild plants was gained through experience in gathering them, and gathering was done with great care. Like wild animals and domestic plants, wild plants were treated with what Gladys Reichard called "incredible respect."[2] Gathering, like hunting, was guided at every step by ritual rules, and just as in the case of animals these rules tended to conserve the species that were gathered.

Indians made apologies to plants they were about to use. "You must ask permission of the plant, or the medicine will not work," said a Navajo, "Plants are alive; you must give them a good talk."[3] There was a prayer for each species of plant, for food plants and medicine plants also. Here again, custom prevented waste. Certain plants could only be gathered by those who had had the proper ceremony, thus further restricting use. Indians thought it worth a long journey over rough country to get the right plants, and even then would collect only as much of each as they needed at the time, so strong were the ingrained teachings that supported conservation. The first plant found of the desired species was usually not picked; instead, an offering was left for it and a song or prayer with the message of human need was given to it to carry to the other plant people. "If you do not give something, you might hurt the plant or the earth."[4] Only the needed part of the plant was taken; in the case of roots, only a portion was removed — preferably a side root — and the plant was replanted with application of sacred pollen and prayers for its continued growth. Sometimes a hole was dug at the side of the plant so it need not be uprooted.

Indians did not needlessly pick flowers. Ruth Underhill remarked that although the Great Plains are at certain times a virtual paradise of colorful wildflowers, the Indians never picked them to add color to their costumes and ceremonials. This was because they looked "on flowers as fellow citizens having as much right on Mother Earth as they had themselves. . . . No Indians . . . picked flowers except for special symbols or for medicine. The wholesale destruction that would have been necessary to decorate a pageant would, perhaps, have horrified them."[5] Navajos love wild flowers, because each one partakes of the spirit of vegetation, and symbolizes all plants. Chants express the wish, "In beauty with abundant plants may I walk."[6] To pluck a flower

50

without need is very dangerous; so none are picked without need. Similar things may be said of other tribes. The Apaches speak of yucca and other plants as their "brothers" and "sisters."[7]

Whenever they could, gatherers handled the plants in ways that would assure their continuing growth. Kwakiutl plant-gatherers left roots in the ground whenever possible, and when too many were pulled up by accident, they put the surplus back in the ground and noticed that it grew again. When roots were pulled, it was done carefully with the offerings of tobacco or pollen, and prayers of thanks. Roots or shoots were often reburied with a prayer for their regeneration. Care was taken not to damage nearby plants of the same species, since they must be members of the same family. When they gathered wild rice, the Menominees always let some of the grains fall back into the water, so that there might be a good crop in the following year.[8] Conservation practices like these may originally have led to the discovery of agriculture, so some students of the subject believe.

The principle that every possible part of anything taken from the natural environment must be used applies in the case of plants as well as animals. The yucca's various parts provided food, fiber for many purposes, and soap. Papagos used every part of the cactus: the stem, woody skeleton, thorns, and the fruit, from which they fermented their cactus wine. The Chumash Indians of California had uses for many parts of plants such as the amole, and tried to keep from throwing away any at all. After an herb had been brewed for medicine tea, it was not simply tossed out, but was buried in the earth with thanks for restored health.

Plants as well as animals were honored in the tribal ceremonials. The Cheyennes celebrated a great *Massaum*, or animal dance, in which several animals were imitated, a ceremony done so "that all the trees and grass and fruits may grow strong." A particularly beautiful dance of the Bella Coola enacted the annual rebirth of vegetation in a manner that was almost allegorical.[9] From beneath an actor representing Mother Nature emerged dancers dressed as willow, aspen, and other plants in the order of their actual appearance in springtime. The South Wind, in ritual combat, repelled the attempt of the North Wind to bring winter back.

Before we go on to speak about trees, we should note that the gathering of minerals followed the same patterns as the gathering of plants. Since Earth is alive, so are the substances that are taken from her body. Potters gathering clay said prayers of thanks and apology to Mother

CHAPTER IV

Earth or Clay Woman, who "gives her flesh."[10] Similar ceremonies were used to propitiate the spirits of turquoise, rock, and salt.

Salt collection usually required a fairly long pilgrimage that took the form of the enactment of sacred legend. The Papagos journeyed across a waterless desert to the Gulf of California, and the Hopis descended into the depths of the Grand Canyon, the traditional place of emergence into this world. Every step of these journeys had symbolic meaning. Several tribes visited a sacred salt lake some distance from Zuñi pueblo. The Zuñis have a tradition that this lake was once much nearer to them, but that the Salt Lady objected to being polluted with trash and debris, and moved the lake. Making amends, the Zuñis "planted their prayer-sticks, praying for forgiveness for being so careless about keeping the lake clean and beautiful."[11]

When metates, baking stones, or grill slabs for frying bread like the *piki* of the Hopis, were cut from a quarry, meal and feather offerings were left, and prayers said to the rock that had given of itself. When the Hopis went to outcroppings to dig coal, or climbed down to Horseshoe Mesa in the Grand Canyon to get the brilliantly blue-colored copper ore, they no doubt used prayers like these and left offerings. The Havasupai had special words of thanks to say when they obtained the red mineral pigment they used as paint and traded to other tribes.

As we would expect from the practices of hunting and gathering, Indians believed that only limited amounts of mineral substances should be taken. Principles of conservation guarded the use of mineral resources in places such as the pipestone quarries of the Plains.

Trees were the most powerful and impressive to Indians of all the plant people, and as we can understand from the words of Tatanga Mani quoted at the beginning of this chapter, they were respected deeply. Of course they were used for many purposes. Trees were of the greatest importance to Indians who lived in forested areas, providing wood for construction, the fashioning of many objects, and firewood for cooking and warmth. Birch bark in the far north and elm bark elsewhere was used for houses, canoes, toboggans and containers of various kinds. The attitudes and practices of hunting applied also to the use of trees. Like animals, trees were seen as having immortal spirits and the power to help or hurt. Their rustling leaves were voices that could speak to the Great Spirit, and if they were harmed unwillingly, they could cry out and seek revenge. Accordingly, when the forest Indians gathered bark, they stripped if off only one side of the tree, so that the tree would not be girdled and killed. Before taking anything from a tree, they expressed their apology and thanks, and often offered tobacco. Dead trees were preferred for firewood, and wood was never wasted, "or all the other trees in the forest would weep, and that would make our hearts sad, too," as the Fox Indians put it.[12] When the Hopis needed cottonwood roots for kachina doll carving, they preferred to gather them in the form of driftwood along stream banks rather than to harm living trees.

The Indians favored trees and shrubs that bore edible fruits or nuts by planting them deliberately near their villages and by sparing them in the forest or when they happened to grow from discarded seeds. Among those so treated were the chestnut, plum, coffee tree, mulberry, and piñon pine.

While all trees were regarded as sacred, some particular species were used in special ways in tribal ceremonies. For example, Pueblo Indians regarded the Douglas fir as a powerful tree, a maker of clouds and bringer of rain. Its Zuñi name means "water comes out arms."[13] When gathering its branches for use in kachina costumes, they chose the suitable tree and climbed as high in it as possible to tie a prayer feather as an offering to the spirits of the mountains who were losing a part of themselves. A Navajo would travel many miles for a pine branch with only one bud, for ritual use. Plains tribes would cut a cedar as the symbolic "tree of the world" and bring it in for a festival with great rejoicing, and similarly treated the forked cottonwood tree that was to serve as the center pole of the Sun Dance. That tree could be cut, or at least the first chop into it made only by a virtuous young maiden who had been chosen for the honor.

In speaking of Indian attitudes and practices involving trees, we must give special attention to the Northwest Coast Indians, since forest resources were of the greatest importance to their way of life. The main wood used was red cedar, but yellow cedar, redwood and alder were also valued. The durable, sweet-smelling, beautiful wood of the red cedar was easily carved with the native tools of horn, bone, shell, beaver incisors, jadeite, and igneous rocks. It is believed that a little iron was traded indirectly from Siberia through Alaska even in pre-contact times. Cedar is so straight-grained that it could be split into long boards with wedges. Other tools used were the stone hammer or maul, adze, chisel, drill, knife, and sander. Trees were felled by driving wedges of elk horn, yew wood, or other resilient materials; fire not being used for this purpose. There were no nails, axes, or saws, a fact

CHAPTER IV

the early European visitors found almost unbelievable when they saw the quality of Northwest Coast carpentry. They were able to make boxes so tight that they were used to store oil. Woodcarving of high skill and decoration also produced dishes, utensils, weapons, tools, masks, and dance rattles. The graceful lines of their canoes were formed by hollowing logs with fire and tools and bending the sides outward by leverage from within. The outside was given a final smoothing with sharkskin. Carvings and painted crests might also be added.

Northwest Coast houses were among the largest and finest dwellings north of Mexico. Rectangular, as large as fifty by a hundred feet or more, they were supported by large carved posts and beams, the walls of cedar boards placed either vertically or horizontally, and the roof made waterproof by overlapping cedar planks. The front might be covered with a painted animal crest design, with an ornately carved entrance post, and partitions inside were elaborately decorated. Inside, the floor was partly excavated to make a large rectangular bench or shelf around the fire pit. Stories were preserved of even greater houses in the past, with multiple levels perhaps resembling some of the wooden architecture of Asia. In the southern region, houses had lower walls and roofs over excavations in the ground. White visitors complained of the fish odors encountered indoors, but it should be noted that Indians couldn't stand the smell of cheese on board European ships, either. Towns comprised of such groups as the Haida were pleasant and healthy, with a row of houses facing the sea, or oriented to catch the southern sun, with the proud crests of many totem poles towering above.

Other parts of trees, such as pliable spruce roots and the soft and easily shredded cedar bark, were used for many purposes including weaving and the manufacture of conical rain hats and capes.

Trees were regarded as sentient beings, sacred in their own right and entitled to great respect from human beings. The Bella Coola said that trees and people once could speak to each other, and although human beings had forgotten the language of trees, trees still understood human speech. Cedar bark was used in most ceremonies as a source of nature's power. Wood was believed to impart an inner life to carvings. And a great tree-pole was considered to be the support of the sky. Thus when Indians were about to cut down a tree, they spoke to it kindly, explaining the use they needed to make of it, and requesting its blessing. Granted the labor involved, they never felled more than they needed. Often only a few boards would be split from a standing tree,

The Plant People

and in such cases, the tree was "begged from;" a prayer was said in words similar to these: "We have come to beg a piece of you today. Please! We hope you will let us have a piece of you."[14] Then a "test hole" would be excavated to see if the wood was suited for boards; if not, the tree would be left standing to make more bark. Its life would not be wasted. Similarly, a gatherer of cedar bark would never strip it more than halfway around the trunk, so as to avoid girdling and killing the tree. A tradition of long standing taught that if the bark-peeler "should peel off all the cedar-bark of a young cedar tree, the young cedar would die, and then another cedar-tree nearby would curse the bark-peeler, so that he would also die. Therefore, the bark-peelers never take all the bark off a young tree."[15] Here is a prayer said by a Kwakiutl woman about to take bark from a young cedar tree:

> Look at me my friend! I come to ask for your dress, for you have come to take pity on us; for there is nothing for which you cannot be used, because it is your way that there is nothing for which we cannot use you, for you are really willing to give us your dress. I came to beg you for this, Long-Life-Maker, for I am going to make a basket for lily-roots out of you. I pray you, friend, not to feel angry with me on account of what I am going to do to you; and I beg you, friend, to tell our friends about what I ask of you. Take care, friend! Keep sickness away from me, so that I may not be killed in sickness or in war. O friend![16]

Since they had such attitudes, and such care in their forest practices, a competent ecologist can conclude that they did no extensive damage to the forest.

This brings us to the question of forest fires. Is it not true that Indians used to start fires in the forest? It is, but their uses of fire must be considered in the light of their careful attitude toward trees. There is no doubt that the Indians used fire to clear fields for planting near their villages, but these fires were carefully controlled. They sometimes drove game with fire, but used backfires and set the fires so as to burn inward toward the center of a circle. When the Navajos had fire drives, they detailed half of the hunting party to keep the fire from spreading outward. Indians sometimes set fire to tracts of forest to clear undergrowth and encourage food plants for deer and game birds. But they never did so indiscriminately or irresponsibly. Indian forest fires were calculated in season, weather, and location to burn over the surface of the ground, not to spread into the forest canopy in destructive

CHAPTER IV

crown fires. Foresters today recognize that fire can be of great aid in certain forest types. If it is carefully used, it can help maintain a wide variety of plant communities.

In this respect, the exact role of fires caused by Indians on the plains is a disputed question. Grasslands are adapted to fire, and may even be maintained by fire. Some ecologists have believed that the Great Plains are essentially treeless because of periodic fires, and perhaps also the actions of the buffalo, but many decades of the absence of both have not resulted in the spread of forests. Indians generally feared prairie fires, and said they did not start them, "as it frightened the buffalo from the country.[17] Buffalo caught in fires could suffer severely, but were often able to escape by running and swimming, both of which they did very well. When Indians used fire to drive buffalo, it was carefully handled. They did not customarily set the grasslands on fire, and when wildfires occurred, they considered them disasters. Indians were not acquainted with the "Smokey the Bear" fetish of modern conservation advertising, but they were always careful with fire. A traveler among the Indians near lake Champlain in 1770 remarked,

> The natives usually make a fire during the night . . . one would think there would be many forest fires . . . but they are very careful to put out the fires.[18]

Fire, of course, was a sacred power, and wise rules governed its treatment. It is interesting to note that in the subarctic forest, and other areas where the vegetation regenerates very slowly after a fire, taking perhaps two or three generations, the Indians strictly avoided setting fires. When the Europeans first landed, eastern America was a land of vigorous forests, not a fire-scarred wasteland. Nevertheless, White settlers ascribed forest fires to Indian incendiaries, and when they rode through an open forest, they would often surmise that Indians had been burning there. But eyewitnesses to Indians starting indiscriminate fires are almost non-existent in the literature. Lightning is a widespread cause of wildfires in North America, and it was easy to blame the Indians for these. Sometimes fires were set deliberately by non-Indians to provide an excuse to fight or remove Indian tribes. For example, when forest fires swept the Colorado rockies in the summer of 1879, newspapers attributed them to the Utes in order to raise the cry, "The Ute must go!" and encourage the seizure of the Ute reservation land for settlers. Open forests are not always the result of fire; they can also be caused by soil type, ecological succession, animal and insect action, topography and climate. The forest Indian was not an agent of deforestation, but a guardian of the forest home.

The Plant People. Tewa women grind corn on metates in the pueblo of Hano, in Arizona, on the Hopi First Mesa. Photographer James Mooney, 1893. COURTESY OF SMITHSONIAN INSTITUTION, NATIONAL ANTHROPOLOGICAL ARCHIVES.

The Plant People. A Pomo Indian man uses a canoe made of reeds to hunt on a lake. Photographer Edward S. Curtis, 1924. COURTESY SMITHSONIAN INSTITUTION, NATIONAL ANTHROPOLOGICAL ARCHIVES.

The Plant People. Mrs. Chua, a Hopi Indian, spreads thin maize batter on a griddle to make piki, *a rolled bread such as seen in the foreground. Photographer J.H. Bratley, 1902.*

COURTESY SMITHSONIAN INSTITUTION, NATIONAL ANTHROPOLOGICAL ARCHIVES.

The Plant People. A Cherokee Indian woman harvests ripe maize. In many forest Indian tribes, women performed the farming duties. Photographer M.R. Harrington, 1908.
COURTESY MUSEUM OF THE AMERICAN INDIAN, HEYE FOUNDATION.

V

All Beings Share the Same Land

THE KIOWAS had an attractive custom that illustrates their attitude toward the environment. When they had taken down their tepees, in preparing to leave a good site where sufficient sweet water, abundant grass, and wood for fuel had been available to them, and where all the people had enjoyed themselves, they recognized that nature had been good to them there, so they left a "gift to the place," an offering of beads, tobacco, or a finely worked leather pouch, in thanks.[1] One cannot help comparing such an act to the thoughtlessness shown by a different kind of "gifts to the place" so often left by non-Indian campers. Indians perceived nature in terms of place. Their relationship was not to nature in the abstract, but to a particular region and to localities within that region. Some of these locations, such as mountains, caves, lakes, and springs, served as especially important points of contact with the spirits or forces whose homes they were. These places became shrines, valuable centers of power. The Indians' relationship to the world was thus structured by a sacred geography. Sacred history was embodied in sacred places where, according to tradition, events of importance had happened.[2] One of the Navajos' monsters of legend, after a fatal meeting with the warrior twins, turned into a great winged rock (Shiprock, New Mexico); while the blood of the giant Yé'iitsoh congealed into the lava field between Acoma and Zuñi. Gobernador Knob (Chi'óol'í'í) was the birthplace of Changing Woman.[3] Weirdly shaped rocks on the canyon walls were seen by Havasupais as the people and animals of their traditions in petrified form. Two huge, erect columns of red rock that stand above the central part of the Havasupai fields are called the Wigaliva, and embody the spirits of the tribe itself. If they ever fall, the Havasupais say, the people will be dispersed. All tribes respect certain place-shrines, where offerings are left, and where piles of stones may rise. The Plains Indians often moved about within their recognized tribal territories, but this did not keep them from a sense of spiritual relationship to the localities that constituted their natural environment. The Sioux, for example honored Bear Butte in the Black Hills as *mato tipi* (Bear Tepee), and brought stones from it as talismans.

N. Scott Momaday, one of the best-known Indian writers today, observed,

> From the time the Indian first set foot upon this continent, he has centered his life in the natural world. He is deeply invested in the earth, committed to it both in his consciousness and in his instinct. To him the sense of place is paramount. Only in reference to the earth can he persist in his true identity.[4]

Sacred geography is never haphazard; the environment is seen as an orderly, balanced system. All tribes regard the four compass directions and upward, downward, and center as particularly holy. These directions are not abstractions, however; they are embodied in the landscape. The tribal home is the central land, guarded in each of the four directions by a sacred mountain that has its own color, animal, bird, and mineral substance, and is the house of a deity. Navajo "sand paintings" often picture the four mountains; in fact, they may often be seen as images of sacred geography and cosmology. The hogan also represents the natural world; its entrance always faces east, the cardinal direction, and in order to serve as the place for a ceremony, it must be round like the universe.

For Pueblo Indians, organization of the natural environment is given a point of reference by the town itself. Most Pueblo groups have a series of migration traditions: the people progressed in an orderly way to various points of the compass until they found the central location, where their village was finally located. The Zuñi term for their own pueblo is *Itiwana*, meaning "middle place." For the Hopi, the center is represented by the sipapuni, the symbolic place of emergence from the underworld, and therefore an aperture of communication. In the form of a hole covered by a sounding board or stone, it marks the center of the town plaza, every kiva, and many fields. The center also can be represented by a spruce tree, as at Isleta Pueblo, or by a pole, as in Taos.

But the functional center of Pueblo thought was the pueblo itself. Around this, the environment was organized in concentric circles, oriented to the four directions, and hallowed by sacred shrines that surrounded the village and provided it with a "protective and life-giving envelope."⁵ The inmost circle was composed of the town and its cultivated fields. Outside this was a circle of hunting territory, of well-known plains, mesas and foothills. Outermost was the sacred circle of mountains, with associated lakes and springs.

For each pueblo, the designation of the sacred mountains was different, depending as it did on the site of the pueblo at the center. Pueblo Indians knew the forms of their sacred mountains well, and had great reverence for them. Models of them could form parts of altars. For some pueblos, lakes were as important as mountains. Blue Lake, at Taos, is a justly celebrated example: "We go there," say the Taos people, "and talk to our Great Spirit in our own language, and talk to Nature and what is going to grow."⁶ The Hopi kachina spirits are believed to live on sacred mountains, especially Nuvatukya'ovi, the House of Snow (San Francisco Peaks, Arizona), but the Zuñi kachinas live in a town under the Lake of Whispering Waters.

Every spirit, and every species of creature, was considered to have a "house" where a shrine could be located. Hopis located the Sun's house on a mesa, the Water Serpents in springs, and the Butterfly House (Polike) on a certain cliff. Ancient kivas have been found on the summits of isolated, cone-shaped hills, and at the mouths of canyons, which are sources of water. Part of the process of initiating youths into the tribe involved teaching them the names of hundreds of sacred localities.

Viewed in the light of a perception of the universe that is basically spiritual, the Indian attitude toward the land itself becomes understandable. The land was the gift of the Great Spirit and the domain of powerful beings. Indians deliberately hallowed their natural environment, perceiving it as a holy, organic unity which surrounded and nurtured them. Jimmie Durham, a Cherokee, gave voice to this characteristic Indian perception of the land in testimony before a Congressional Committee:

> In *Ani Yonwiyah*, the language of my people, there is a word for land: *Eloheh*. This same word also means history, culture and religion. This is because we Cherokees cannot separate our place on earth from our lives on it, nor from our vision and our meaning as a people. From childhood we are taught that the animals and even the trees and plants that we share a place with are our brothers and sisters.

So when we speak of land, we are not speaking of property, territory or even a piece of ground upon which our houses sit and our crops are grown. We are speaking of something truly sacred.[7]

Land in the Indian view was not "owned" in the sense that word had in Western European societies; rather it was held in common. In the widest meaning, Indians felt that all living beings share the land, and that includes plants and animals as well as human beings. But in a more local sense, ownership was tribal and the land was considered to belong to the community even when it was used by families or individuals. Among Indians, cooperation and group interests predominated, particularly where ecological conditions meant subsistence living in a difficult environment. They were tolerant of individual desires, appreciative of individual contributions to the group, and slow to use sanctions against individuals, but they were not individualists. An Indian felt himself or herself primarily as part of the family, clan or tribe, and the world of life. Most ceremonies and economic activities were done by cooperative groups. A competitive attitude was regarded as antisocial or malevolent. Generosity, hospitality, and the customary exchange of gifts provided for sharing food and goods throughout the local group. As we noted in the chapter "The Powerful Animals," a hunter was expected to divide the kill among others. For this he might well receive repayment in the form of cornmeal, squash, or a meal. After the Pueblo harvest, those who had had a poor crop were allowed to glean in the fields, and delegated chiefs took care of extra stores of grain that they could give out in case of need.

This cooperative ethics extended also to land use. "This beautiful land that is our mother,"[8] the sacred resting place of the ancestors and the topography of holy shrines, was not something to be owned privately. "We do not own land, we simply use it,"[9] was a characteristic Indian statement. Such use was assigned by the group (tribe, village, clan, etc.) in a customary, accepted manner. Boundary lines were set in agricultural lands from time to time and then hallowed by long use. A family had a definite block of land in which to plant, and felt entitled to use it, but recognized that it was assigned to them by the group, not their exclusive property. Land disputes were not unknown, but were usually settled peacefully.

CHAPTER V

Along the Northwest Coast, the feeling about property was different from that found in other parts of North America, and their sense of land ownership was strong. Villages and the people who inhabited them took geographical names derived from their location. The use of definite, recognized resource territories for fishing and hunting was the right of family groups, and that right descended in lineages, just as names and crests were inherited. Each family had its own stretches of beach, and whatever drifted ashore there was theirs to use or bestow as a gift on others. Sections of streams for fishing, smoke-house sites, berry patches, and tracts of forest for hunting, were also family territories, and treated in a way analogous to sacred precincts; removal of anything without permission could be punishable by death. Wars between groups were sometimes fought for possession of these resource lands, and the successful side drove out the losers, but such incidents were relatively unusual. Raids for captives and loot were more common. The possession of a definite territory on which they would always be dependent for food and other products made conservation a wise and advantageous course for each family. Salmon almost always return to exactly the same stream where they were spawned, and a tree spared could be found and used the next year.

Differences in the way they viewed and treated the land became a serious point of conflict as soon as Europeans and their American descendants began to move into the regions formerly inhabited only by Indians. The newcomers thought the Indians had no proper sense of the pride of ownership or use of the land, since they had no deeds to property, often moved from place to place, and did not fence or plow the land. Similarly, Indians thought that their new neighbors lacked love for the earth because they tried to cut it up, buying and selling pieces of it, wounding it by plowing, and moving on when they had taken what they wanted from it. Chief Seattle said, "[my people] still love these solitudes . . . And when the last Red man shall have perished from the earth and his memory among white man shall have become a myth, these shores shall swarm with the invisible dead of my tribe."[10] The Indians' ties to the land were those of kinship and ritual, not those of the ownership of property and proprietorship. Indians did own the land in every recognized sense of occupation and use, of course, but as a fabric woven of spiritual entities with whom the Indians maintained an extended series of intertwined relationships, it could not be sold, and it was held in common by the tribe. An Indian might say or think, "This is our land," seldom "This is my land." The Iroquois said, "The soil of the earth from one end of the land to the other is the property of the people who inhabit it."[11] Indians, as we have seen, had a strong sense of place. For them, space consisted of meaningful locations, just as time consisted of meaningful events. When Chief Joseph wanted to indicate what might now be called a "point in time," he specified instead a location, "From where the sun stands . . ."[12] He saw neither time nor space as an abstract quantity that can "run out." The land belonged to the past and future; it entombed the bones of ancestors, which added to its sacred nature, and it would be the home of the children and their children after them. Common ownership did not extend beyond the tribe, although in the later struggle against the appropriation of Indian land by the White, Tecumseh and a few others had the vision of a whole continent given by the Great Spirit to all Indians, and epitomized the Indian feeling for the land in the famous words, "Sell the earth? Why not sell the air, the clouds, the great sea?"[13]

The most serious blow to Indian cultures was, of course, the appropriation and alteration of their land by the non-Indian invaders. At first the Indians did not realize what was happening, since many of the early treaties guaranteed Indian rights to hunt and fish on the land that had been sold, and these were important signs of ownership to Indians, who had no concept of land as a salable commodity subject to exclusive private ownership.

Indian love of the land persisted after the European and American conquests. Taos Pueblo youths continued to be taught the names of sacred places that had become the property of others, or had been destroyed. When asked about this, they maintained, "We own them in our soul."[14] Hopi Indians still claim their traditional land (although some believe it is wrong to fight for it as a possession in the United States courts), since it was used by them from time immemorial and holds their sacred places. Navajos who were forced to live in a "concentration camp" at Fort Sumner during the Civil War lamented deeply the loss of their own land (which later was restored to them) in terms like these:

> Oh beloved Chinle,
> That in the springtime used to be so pleasant!
> Oh beloved Black Mountain!
> Would that one were at these so-named places![15]

Today, most Navajos feel that the place for them is their homeland between the sacred mountains, and that if they try to live outside it, they

CHAPTER V

will fall out of harmony with nature and the gods. To this they attribute difficulties experienced by numbers of the Navajos who have moved to distant cities like Denver and Los Angeles.

As we noted in the first chapter, the result of the Indian attitude toward land in the years before contact with Europeans was the practicing of life styles that were not destructive. There were no great areas of depleted game, no major deforestations, and no noticeable pollution except for a few tree-covered trash piles. The Indians used their natural environment without ruthlessly exploiting it. It is unfortunate that the same cannot be said about the later owners of the American land. In the Southwest, for example, the earliest Spanish and American explorers remarked on the luxuriant grass and abundant game of the region, two things now noted by their absence. Tecumseh noted that the non-Indians had a different attitude to the air and sea as well as the land. The Indian elders knew that all beings share the same air, just as they could not understand why others did not see it clearly, as they did.

Hunting — the holy occupation. A Cree Indian hunter demonstrates moose-calling with a bark horn. Photographer Edward S. Curtis, circa 1927. COURTESY SMITHSONIAN INSTITUTION, NATIONAL ANTHROPOLOGICAL ARCHIVES

A boatload of Makah Indian men brings a whale ashore at Neah Bay, Washington. Sealskin floats are nearby in the water. Photo circa 1926. COURTESY SMITHSONIAN INSTITUTION NATIONAL ANTHROPOLOGICAL ARCHIVES.

Hunting – the holy occupation. Ojibwa Indians in a birch bark canoe, hunting with bow and arrows. Photographer Roland W. Reed. Photo circa 1900. COURTESY SMITHSONIAN INSTITUTION NATIONAL ANTHROPOLOGICAL ARCHIVES.

Hunting — the holy occupation. Ojibwa Indians, equipped with snowshoes and rifle, set out to hunt deer. Snowshoes were invented by the forest Indians. Photographer John K. Hillers.

COURTESY MUSEUM OF THE AMERICAN INDIAN, HEYE FOUNDATION.

VI

The Gifts of Mother Earth

A Hopi MEDALLION shows a circle shared by two plants, maize (Indian corn) and beans, that grow up together through the center and complete the two halves of the design with their leaves. The artist explained the design by saying that beans and maize have a spiritual unity, growing best together. Other tribes that practiced agriculture used similar designs, because they often planted these vegetables, two of the most important Indian food crops, side by side. The cornstalks provided a climbing pole for the beans, and the beans, legumes whose roots contain nitrogen-fixing bacteria, provided needed elements to the soil for the maize. The planters had discovered and encouraged a natural partial symbiosis, a relationship between two species in which each provides something the other can use. This was only one of the ways Indians knew to preserve the fertility of the soil and to assure that fields could be used to grow crops for many years.

Through long experience and deep reverence for Mother Earth and the growing things that they regarded as her children, Indians learned to apply to farming what we would call the principles of ecology. As one Papago said, in farming "everything had to be taken care of."[1] This care was regarded by the Indians not as a matter of economic self-interest, but as a sacred way of life. "Agriculture is a holy occupation," said Pete Price, a Navajo, quite unaware that exactly the same words could be used to describe hunting in the Northeast, "Even before you start to plant, you sing songs. You continue this during the whole time your crops are growing. You cannot help but feel that you are in a holy place when you go through your fields and they are doing well."[2] Maize plants were seen as holy people, and were raised with the care one would give a baby. Agricultural experience was embodied in sacred tradition and ceremonial.

Not all tribes practiced agriculture, and among those who did, there were many differences, so we will discuss examples from three different

CHAPTER VI

cultural areas. First we will look at the Eastern forest Indians, who both farmed and hunted. Among these people, caring for the crops was women's special task. Women owned the farmland and its produce, which was only one of their important roles in the tribe. They knew how to encourage the earth to bear fruit, as they knew how to bear children. The blood payment for a woman was often twice that for a man,[3] and among the Iroquois it was the women who elected the tribal leaders. Their work was not drudgery; it carried honor.

The most important crops were maize, beans, and squash, which might be cooked fresh, or dried and saved until needed. Maize was ground on shallow mortars with hand-stones, and hollowed vertical tree-trunk mortars with long cylindrical wooden pestles were also used.

The practice of burning the fields in the autumn returned some of the elements needed by the plants to the ground. After a number of years in one place, the forest Indians would clear a new section of land, possibly moving their village as well. This technique, called "slash and burn" by some authors, allowed the forest to reclaim the old fields and restore their fertility over the generations. Traditional agriculture did not destroy the soil or encourage severe erosion. Forest Indians knew the use of fish as fertilizer. One of the best known stories among American schoolchildren is that of Squanto, the friendly Massachusetts Indian who taught the Pilgrims how to bury fish in their corn hills so their crops would flourish. Skeptics have pointed out that Squanto had been to Europe before that, and might have learned about fertilizing there, but it appears that the practice was indeed of long standing among the coastal forest Indians, because *menhaden*, the Algonquian word for a fish used commonly in this way, means "fertilizer" in that language.[4]

Tobacco was valued for ceremonial purposes, and much trading was done for its seeds. Since tobacco was used by the Indian in ritual, not as a habitual indulgence, it did not produce the general health problem it presents in society today.

The ceremonies used by hunters were adopted into agricultural activities as well. Festivals with communal dances and sacred songs followed the important events of the agricultural year. The first fruits were especially honored, and Thanksgiving as a religious feast among the forest Indians was celebrated long before 1620, in gratitude for the nourishing gifts of Mother Earth.

For a second example, we can go to the eastern half of the Great Plains, where some of the tribes practiced maize agriculture, although it was not as important a part of their economy as it was for the Pueblos, the Southeastern tribes or even the Northeastern Forest Indians. They remained buffalo-hunting Indians who supplemented their diet with crops. Maize was not native to the plains, and was raised only with great difficulty in the higher areas due to fluctuations in weather patterns from year to year. The Plains Indians planted many hardy strains of maize, fast-growing and resistant to frost and drought. They also grew beans and squash, and were quite possibly the original domesticators of the sunflower. Agriculture was the task of women among the Plains Indians, as it was among the Forest Indians, and that is probably a result of the fact that most of the tribes on the Plains had come from forested areas in earlier times. Tobacco was an exception; because of its ceremonial use in the sacred pipe, it was tended by men. It was raised by old men even among some tribes that did not otherwise plant crops. Tobacco was permitted only to old men and a few old women; young men were taught not to use it, because it shortened the wind they would need for physical exertion.

Agricultural tools were simple implements like the digging stick, bone hoe with wooden handle, and antler rake. Like Indians in other areas, they objected to scarring Mother Earth's breast with plows.

Crop rotation was unknown. Brush and dead plants were burned before the spring planting, and the practice of alternating rows of corn and beans was used, but other means of fertilizing the soil were absent. In fact, the women carefully removed horse manure from the fields, because they noticed that it introduced alien weed seeds and harbored insects. There was no such objection to buffalo chips, although buffalo were wary enough to avoid Indian gardens, and any chips found would doubtless have been burned as fuel. The maize crop was watched constantly by young women as it matured, to keep birds and squirrels away and give warning of raids.

The ripe maize was treated with the greatest respect, called "mother" and associated closely with Mother Earth. No metal knives were allowed to touch the cornstalks by some tribes, nor were any kernels left scattered on the ground. The harvested corn was stored carefully in underground cache pits that had been blessed ritually. Seed corn was consecrated and strictly protected.

Maize was sacred; its cult among the corn-raising tribes was recognized as approaching the buffalo ceremonials in importance, and closely associated or even integrated with them. As Black Elk pointed out,

CHAPTER VI

> The corn that we Sioux now have really belongs to the Ree (Arikara), for they cherish it and regard it as sacred, in the same manner that we regard our pipe; for they, too, have received their corn through a vision from the Great Spirit.[5]

The sacred bundles of the Arikara contained ears of corn, as those of the Sioux had the pipe given by and symbolizing the buffalo. Corn figured in many incidents of sacred history among the planters, and there were sacred songs that served as the Omahas expressed it, to "sing up" the corn. The Mandans had a Corn Priest who wore paint in special patterns and officiated at rites throughout the growing season.[6]

The third cultural area that will serve as an example of ecological attitudes and practices in Indian agriculture is the Southwest. In this land of little rain, it seems paradoxical that agriculture should have become a dominant activity. Yet the fact that it did emphasizes the ecological problems that Southwestern Indians had to deal with, and justifies a more detailed discussion. Crops provided probably 85 percent of the food consumed by the Pueblo Indians, and much of their clothing and other needs.[7] The Pueblos constituted the economic and cultural center of gravity of the entire Southwest. The Pimas, Papagos, some Yumans including the Havasupai, and many Navajos were predominantly agricultural people, getting 50% - 60% of their food from domestic plants.[8] Even a few members of those tribes that lived by hunting and gathering, like the Apaches, might plant a field of corn in a favored spot and return later to harvest it. Maize was not unknown to any Southwestern Indian tribe.

Careful treatment of the earth and arduous methods of conservation were necessary for Indians who depended on agriculture in a land of erratic water supply, short growing seasons, winds and often shallow, unstable soil. While the climate might have been better when agriculture was introduced — corn from Bat Cave, New Mexico, has been dated at 3000 B.C. or earlier[9] — the Southwestern farmers survived periods of climate even worse than at present, though not without serious setbacks.

The major crop, as will be expected from what we have already said, was maize, the crop most Americans call "corn." The Pueblo Indians distinguished at least nine different kinds, of almost as many different colors, and used careful seed selection to keep the desired characteristics pure. The Hopi have developed varieties of maize that are admirably adapted to the conditions in which they have to grow them. Hopi corn grows rapidly, reaching maturity in 115 to 130 days, but it is not tall — its bushy form well resists the dry, tearing desert winds. (Iowa corn would be knocked flat by an Arizona wind.) In addition, it sends up a long shoot underground before the leaves appear, enabling the seed to be planted a foot or so deep where some moisture persists through the dry season of spring, and its first root is single and deep-thrusting, unlike the bunch of short roots produced by other varieties of maize. Piman (Pima and Papago) corn is well adapted to the hot lowlands; its ears are smaller than Pueblo corn and most of it is white or yellow.

Beans constituted the second staple crop of Southwestern farmers. They grew several varieties, including kidney and lima beans, and tepary beans of Piman origin. A third category of cultivated plants included squashes, gourds, and pumpkins. The sunflower, a plant domesticated in North America, was grown for its seeds. Pueblos and Pimans planted New World cotton (*Gossypium hopi*) and wove it into clothing. Hopi cotton has very fine fibers, and like Hopi corn, is adapted to a short growing season by exceedingly rapid growth. Most tribes also planted tobacco for ceremonial use. Several species of wild tobacco grow in the Southwest, but they also traded for seeds from other areas. Seeds of useful wild plants were collected and sown in Southwestern Indian gardens, including mint, bee plant, walnuts, wild potatoes, and devil's claw (the latter used in basketry and ceremonial costumes).

The usual agricultural tools were simple implements like the cylindrical or paddle-shaped digging stick, and a short weeding hoe (pushed forward in a kneeling position). They had no plows, since they had no animals to pull them, but even after the introduction of European agriculture, the Hopis and others refused to plow because of the dangers of wind erosion and increased evaporation of precious soil moisture.

Men did almost all outdoor agricultural work in the Southwest, but among the Pueblos, women owned the land, the seed and consequently the stored crops; there it was women who practiced seed selection, and they customarily thanked the men every time they returned home after a day of work in the fields.

Fields were relatively small, and had to be located carefully, taking into account the water supply, danger of frost, and soil. Soils were judged by observing color, taste, texture, moisture content, and by noting the species and vigor of the wild plants that grew on them.

CHAPTER VI

Southwestern Indians developed many methods of water collection, diversion and irrigation to conserve water supply and conduct it to the fields while preventing erosion as much as possible. Floodwater irrigation, using the erratic rainfall and the water briefly carried by intermittent arroyos, was practiced by the Pueblos, Navajos, and Papagos. Colorado River Yumans depended on river floods. Ditch irrigation from permanent streams or rivers was used by Pimas, Havasupais, and some Pueblos. Some tribes also irrigated small gardens from permanent springs.

Floodwater farming is a widespread but difficult method well described by Kirk Bryan:

> The areas utilized are variable in size and location, but each is chosen so that the local rainfall may be reinforced by the overflow of water derived from higher ground. The selection of a field involves an intimate knowledge of local conditions. The field must be flooded, but the sheet of water must not attain such velocity as to wash out the crop nor carry such a load of detritus as to bury the growing plants. Such conditions require a nice balance of forces that occur only under special conditions. Shrewd observations and good judgment are necessary in the selection of fields.[10]

Such locations are usually at the bases of mesas or mountains, on alluvial soil at the mouth of arroyos. The cooperation of the forces of nature is absolutely essential.

Floodwater farmers have to contend with the strongly localized nature of summer rainfall in the Southwest. A thunderstorm may soak one field (or pelt it with hail) and not even touch another only a few feet away. So Indians built collection works: retention dams, ditches and dikes, sometimes extending for miles, providing runoff for the fields in which they converged. Extensive systems have been found at Chaco Canyon and Mesa Verde, and their purpose is quite evident. By artificially augmenting the area of watershed, they increased the chance that rains would fall in at least part of it, and save crops that otherwise might have perished from drought.

Countless works were built, maintained, and rebuilt to prevent or retard erosion. Soil and rocks were loosened with digging sticks and carried in baskets. Ditches were carefully engineered to be shallow but wide, to prevent the formation of an erosion channel. It has been noted that their usual cross section (thirty feet wide by one foot deep) matches closely one recommended by the Soil Conservation Service for a vegetated channel; maize might even have been planted in some of the ditches.[11] Such ditches are sometimes hard to see on the ground today, though they often appear clearly in aerial photographs. Earth, rock and brush dams were built across arroyos to slow or divert floodwaters and catch soil, and every effort was made to distribute water evenly across the fields through leveling, building brush and soil barriers, and trenching. Zuñi fields were often surrounded by a high ridge of soil to aid in retaining floodwater. When conditions were right, the soil would be enriched annually by a new deposit. Terraces were built on slopes, ranging from simple boulder benches to elaborate step-like constructions with walls, and sometimes utilizing organically rich refuse from settlements.

Wind is second only to water as a force of erosion in the Southwest, and brush windbreaks were kept in the field to retain soil. The ground was broken as little as possible, and stumps of cornstalks were not removed. A badly wind-eroded field might be left fallow until bushes grew and collected enough windborne soil to allow clearing and replanting. Indians tried not to let erosion get out of hand; as the Papagos reported, in early times, "if there were signs of erosion in a field, immediate steps were taken to stop it."[12] All these things could be done only by constant cooperative labor in the right season. No project was undertaken without careful consultation by the people concerned, and no "cutting" was made into Mother Earth without ceremonial apology.

Irrigation by canals from permanent rivers and streams was practiced at several places in the Southwest. Ancient clay-lined canals of the Hohokam in the Gila Valley constitute the most extensive pre-Columbian irrigation system in North America, covering more than 100,000 acres. They show evidence of frequent maintenance and redigging. The same methods were continued into modern times by the Pimas, using the waters of the Gila and its tributaries. Construction of canals and river dams (of pilings, brush, logs and rocks) was undertaken by men from several villages working together. The irrigated land was carefully leveled. Some of the Rio Grande Pueblos used similar techniques on a smaller scale. Spanish explorers saw irrigated fields with canals near Acoma in 1582.[13] The Havasupai have long irrigated their small canyon floor, less than a square mile in area, with water conducted in ditches from a never-failing stream.

In a dry climate, irrigation brings with it the danger of salinization as water evaporates and leaves its salt content concentrated in the soil. Indians of many Southwestern tribes could recognize this problem, which also occurs naturally, by observing the white crust, salty taste of the soil, the growth of salt-loving plants, and the changed consistency of the soil, which grew sticky when wet and made a creaking sound when they walked on it. Pimas tried to counter salinization by throughly flooding such soils before planting in them. Hopis planted in well-drained areas and incorporated crop residues into the soil.[14] Crop rotation and manuring were unknown in the Southwest, but burning for fertilization, fallowing, and the practice of planting corn together with nitrogen-fixing beans were known, as we have seen.

The year's round of agricultural activities took place in harmony with the cycles of the natural environment, and Indian farmers had an intimate knowledge of these cycles. Since little rain falls in spring, when it would be most useful, the floodwater farmer had a dilemma: plant early, and take the chance that the growing plants will dry out before the rains begin, or plant late, and risk an early frost before they have matured. Hopis plant in spring and count on the previous winter's moisture to carry the young plants through June; while Papagos, where the growing season is longer, wait to plant until the summer rains begin. Their riverside neighbors, the Pimas, plant at the beginning of the frost-free season, and sometimes manage to get two successive harvests from the same plot of ground.

All tribes kept careful track of the changing seasons and could judge when to perform each activity from what they observed in nature. The Navajos, for example, planted when the aspen or wild cherry leafed, or when the yucca blossomed. The Pimas waited for the cottonwood and mesquite leaves and the return of the white-winged dove. The Hopis and Zuñis had ceremonial officers who watched the sunrise each morning from a fixed spot, made offerings and prayers, and kept the calendar by observing the successive points on the horizon where the sun, that great source of life, appeared. Their observations were reported to the people by a town crier. The Navajos in Canyon de Chelly observed the shortening shadows of red rock walls. Many tribes watched the positions of stars as well, particularly the Pleiades. The moon's phases were known, and the months or "moons" had names (at least during the growing season), based on the occurrence of changes in nature, agricultural activities, and sacred ceremonies of the solar calendar. It has been noted that members of the same tribe living at different elevations might not agree on what "moon" was currently in progress, it being more important to keep in step with the local environment than to mark abstract time. Indians watched the sky and observed the actions of birds and animals to predict changes in the weather: rain, snow and sandstorm.

All Indians treated their crops with extreme care, and made minute observations of the process of growth. Hopis distinguished nine stages in growing maize, from planting to harvest; while the meticulous Navajos distinguished 20 stages for maize, 13 for melons, and nine for beans. Seeds were guarded carefully, and surrounded by ritual, as sacred givers of life. "Do we not live on corn, just as the child draws life from the mother?"[15] remarked a Hopi from Oraibi.

After the fields had been cleansed, literally and ritually, planting was done carefully in hills, with hills in one row matching spaces in the next row, so as to give each hill the greatest possible room. This year's hills went between last year's hills. The Navajos often planted spirally, following the sun. Several seeds were put in each hill, deep in a hole made by the planting stick. Hopi planters were expected never to throw anything away while in the fields. It was usually a group effort, preceded by prayer accompanied by song, and done according to traditional principles.

Through the growing season, weeds were cautiously removed to prevent them from taking up needed moisture, and the soil loosened around the plants with care not to damage them; as this Pima example shows:

> If a man accidentally cut off a stalk of corn while hoeing, it was buried in the field, or tucked under his belt and carried home at night. He felt very badly about the experience for corn was his very life. It was just like "killing his own life" or injuring a friend.[16]

Careful thinning could be done, of course. Brushwood and stones were placed so as to protect growing plants, mice and prairie dogs were trapped, and while the ears of maize were actually ripening, a guard was posted in the fields to keep away strangers, birds and animals. Harvest was a time of hard but cooperative work accompanied by songs, dances and races.

All Southwestern Indian farmers believed that ritual was just as important for the growing crops as physical work in the fields. "Singing up the corn" was the way they described it.[17] Every stage in the development of maize had an appropriate song. Each tribe had agricultural dances and songs, often with costumed dancers. Papago harvest dancers imitated clouds, winds, birds, animals, plants, sun and moon, etc. The Navajos had songs and ceremonies, even the better known healing chants, may be said to be for the harmony of the universe and the prosperity of the people, and therefore for good weather and crops as well.

Planting had its own rituals. Races were run to encourage the plants to grow rapidly, the clouds to hurry near, and the sun to have strength in his summer course. Tobacco was smoked to bring clouds, water sprinkled to call rain. Hopi women dashed a dipper of water over their men on the day of the planting. Maize was often planted in the warm kivas (sacred chambers) where it shot up early. Prayers were said. Here is a Havasupai prayer, said by the planter as he faces a painted symbol on the canyon wall that marks the home of the patron spirit of maize:

> I like my crop to be good corn. I don't want it to go dry, or have small ears on the corn, so You, the Corn up on the wall, do look over my crop and help me along where we all have a good crop this fall.[18]

There were ceremonies before communal work on irrigation ditches, ceremonies against grasshoppers, and great harvest festivals.

Maize was recognized in the Southwest as the sustainer of human life. Its pollen and ground meal, most sacred of substances, appear as offerings in almost every ceremonial, sprinkled or used to outline symbolic designs. The Apaches, who had few cultivated plants, used wild cattail pollen instead of maize pollen. Ears of maize were powerful talismans. Among the Zuñis, the newborn infant was presented with an ear of maize, receiving a "corn name." An ear of maize was put in the place of death as the "heart of the deceased" and later used as seed corn. "Corn is the same as a human being, only it is holier,"[19] said a Navajo.

Tobacco, too, was a sacred substance, its smoke identified with clouds and moisture, and fragmented leaves or smoke offered in ceremonies. In some tribes, its culture was a guarded secret surrounded by complex ritual. Its use by young men was discouraged, and it was almost never smoked outside of ceremonies.

Prayer sticks were relatively short twigs taken from living plants, decorated with feathers and paint, and used as visible signs of the peoples' prayers. Placed at altars, shrines, springs, or at any other locality of special concern, they were believed to be one way in which human beings could make a return for the good things received from the natural environment.

In the dry Southwest, where rainfall varies exceedingly from year to year, most agricultural ceremonies were intended to be prayers for rain. Clouds and everything else connected with rain appear as ritual motifs, for ritual was deemed essential in bringing moisture to a thirsty land. Not that the ceremonies were believed to be pure magic; Indians were "wise enough not to hold a ceremony that was against the order of nature." Ruth Underhill once asked a Papago why he would not pray to "pull down the clouds" in April (the Yellow Moon), when dry weather prevents any planting, and received a reasonable reply: "There are no clouds in the Yellow Moon."[20]

Rain was, however, a valid objective of prayer in the Indians' view, since it was sent by powerful beings (sometimes four in number, one for each direction, like the Chacs of the Central American Mayas and the Tlalocs of the Aztecs in Mexico. A widespread belief held that clouds were the spirits of the ancestral dead, who could be implored to return, bringing rain to their needy descendants. A Hopi might say to someone who had recently died, "You are no longer a Hopi, you are Cloud."[21]

Several methods of calling rain were used. It was felt that if the ceremonies were done sincerely and without error, rain must come. Vivid descriptions of rain and the use of words associated with water were believed to produce the results they called to mind. During the Navajo rain ceremony, all conversation had to be framed in words connected with moisture. A Papago rain song is simple but evocative:

> At the edge of the mountain
> A cloud hangs,
> And there my heart, my heart, my heart
> Hangs with it.
>
> At the edge of the mountain
> A cloud trembles
> And there my heart, my heart, my heart
> Trembles with it.[22]

Another principle observed in rain ceremonies is that of sympathetic magic: like produces like. Water may actually be sprinkled or poured. Hopis often buried a jar full of water in the center of a cornfield. Plants that grow near springs, or at higher elevations where it rains more often, appeared in rituals. Smoke and yucca suds looked like clouds, and were believed to attract the thing they resembled. The Papagos called rain by preparing and drinking cactus wine. A common Papago

CHAPTER VI

exhortation might well be translated as follows, using a modern expression that captures the idea very well:

> Drink, friend! Get beautifully "soaked!"
> Hither bring the wind and the clouds.[23]

Ceremonial instruments, including rattles, drums, the bull-roarer, the morache (a notched stick with gourd resonator stroked with a deer scapula) and the wooden lightning frame, recalled in form and sound the thunder, wind and lightning, asking them to come, and the rain with them.

Offerings were intended to compel the spirit beings to send rain. A Zia Pueblo rain prayer runs as follows:

> I send you prayer sticks and pay you sticks of office, kicksticks, hoops, shell mixture, various foods, that you may be pleased and have all things to wear and eat. I pay you these that you will beseech the cloud chiefs to send their people to water the earth that she may be fruitful and give to all people abundance of all food.[24]

Indians also appealed to the sympathy of the powerful spirits, as in this Papago prayer to the Rain Being, asking him to have pity on living creatures:

> The trees you planted have no leaves;
> The birds you threw into the air
> Wretchedly flit therein and do not sing;
> The beasts that run upon the earth
> At the tree roots go digging holes
> And make no sound . . .[25]

People who accept the modern scientific view probably will not believe that these ceremonies brought or increased the rain, but they cannot deny that they gave expression to the needs and earnest desires of the people who performed them. And one remembers that modern technology can not yet bring rain with any degree of certainty.

Wherever Indians farmed, their methods conserved both water and land, greatly reduced the danger of erosion, and were "delicately adjusted to the needs of the early communities."[26] To succeed, they needed to provide constant vigilance and difficult labor at the proper seasons. Their practice was motivated by a religious view of nature. Their ecological understanding was embodied and preserved in traditions and ceremonies that included stories and songs. And when they succeeded, they could survey their handiwork and say, as the Governor of Zuñi once did, "Zuñi farming always keeps the land good."[27]

Intensive horticulture, using methods that prevent erosion, are exemplified by these terraced fields that are irrigated by a spring. Situated near the Hopi pueblo of Walpi in Arizona, the actual location is called "Weepo (Cattail-Water) Gardens." Photographer Jesse H. Bratley, circa 1902.

COURTESY SMITHSONIAN INSTITUTION, NATIONAL ANTHROPOLOGICAL ARCHIVES.

The Plant People. A Navajo cornfield and summer brush shelter near Holbrook, AZ, contains the low-growing variety of maize that is well adapted to desert conditions. Photographer F.A. Ames, 1889.

COURTESY SMITHSONIAN INSTITUTION, NATIONAL ARCHIVES.

A Cheyenne Indian woman hangs meat on racks to dry. Using this method, meat could be stored for long periods of time, thus also contributing to the conservation of buffalo herds. Photographer Edward S. Curtis, 1927.

COURTESY SMITHSONIAN INSTITUTION, NATIONAL ANTHROPOLOGICAL ARCHIVES.

The avoidance of waste. Many salmon hang to dry on racks (background) made by the Coast Salish Indians, Fraser River, British Columbia. Fish caught during the salmon run could thus be preserved for use during the rest of the year.

COURTESY MUSEUM OF THE AMERICAN INDIAN, HEYE FOUNDATION.

VII

The Wisdom of the Elders

WHEN AN OREGON TRIBE gathered around the campfire after dusk on a long winter night, the young men often heard one of the elders tell a story like the one that follows:

"Once upon a time, a boy went on a quest for a guardian spirit, and after a long and difficult search he met a powerful being named Great Elk. Great Elk gave the youth the skill to be a successful hunter, while warning him never to kill more game than he needed at one time. But the young hunter soon found he could slay as many animals as he wanted, and he violated Great Elk's command by slaughtering whole herds of wildlife, even including elk. Seeing this, the guardian spirit withdrew his protection, causing the young man to lose his hunting ability and starve to death."[1]

Those who heard that story would have had the opportunity to learn both an ecological principle and a tribal attitude toward the natural world. This was a way in which wisdom was communicated through the generations in an Indian tribe. The elders had great experience and were honored for it; their knowledge was passed down to the young in the form of impressive, interesting and often humorous traditions.

The elders were skilled hunters and planters, knowing the habits of beasts and birds and the appearance and ways of growth of plants. Indians learned ecological principles and relationships from nature in a very direct way. Hunters might spend days or weeks alone in the forest, observing and interacting with animals and plants, land, water and weather. Living with nature at first hand, they would soon see the results of carelessness or waste.

So sensitively aware of things in nature were the Indians, and so able to observe the signs and habits of animals and plants, that early frontiersmen sometimes thought they had a mysterious "sixth sense." Actually, the Indians' cultural training and experience of the forest had taught them to perceive things that others would miss. Since they lived in direct dependence on the natural environment, the Indians had detailed knowledge of its workings. They developed an extensive body of knowledge about nature that amounted to an ethnic science. Indians were keenly observant and rational, but would make explanations that would be excluded even as hypotheses by modern Western science, because they were often subjective and mystical. But they were always based on empirical observation and experience.

They had names for virtually every visible species of animal and plant that would be recognized by a modern taxonomist. Navajo herbalists could name and describe over 300 different plants, and had uses for most of them in medicines, foods, arts and ceremonies. Small distinctions between species were noted. The Tewa, for example, have a different name for each kind of cone-bearing tree in their homeland. It has been said that every Hopi is a practical botanist. They knew the characteristics of hundreds of plants, and the effects they might have on the human body through the power in them, as they would explain it. They knew that roots take up moisture, and that leaves provide food for a plant.

They took stock of their environment, knowing how much there was of each species and where it was to be found. Hunting was governed by an accurate census of animals and a knowledge of how many could be taken without depleting the supply. They knew the breeding habits of animals and their needs for food and shelter. Some Indian stories show a knowledge of food chains and the balance of nature. The flight, migration and nesting of birds was familiar to them. The screaming of owls at night and the flight of crows and geese were to them warnings of changes in the weather. The Indian names for the "moons," or months, show at least a part of their detailed knowledge of the seasonal cycles and rhythms of nature: when flowers or fruits would appear, when the young of animals would be born, when the lakes would freeze, when the birds would return. The thunder being was identified as a bird because the season of thunderstorms began at the same time that the birds returned from the south. The Indians' science was a blend of observation, reason, insight and nature mysticism.

Indians did not consider that they themselves had discovered the uses of things in nature, rather that these uses had been revealed to them by animals and spirits: "The buffalo is wise in many things, and thus, we should learn from him and should always be as a relative with him."[2]

CHAPTER VII

Respect for the wisdom of animals came from direct observation of their behavior. As Ohiyesa (Charles Eastman), a Dakota elder, said, the Indian "had faith in [the animal's] instincts, as in a mysterious wisdom given from above."[3] This was also well expressed by Brave Buffalo, a Teton Sioux:

> I have noticed in my life that all men have a liking for some special animal, tree, plant, or spot of earth. If men would pay more attention to these preferences and seek what is best to do in order to make themselves worthy of that toward which they are so attracted, they might have dreams which would purify their lives. Let a man decide upon his favorite animal and make a study of it, learning its innocent ways. Let him understand its sounds and motions. The animals want to communicate with man, but Wakantanka does not intend they shall do so directly — man must do the greater part in securing an understanding.[4]

Luther Standing Bear agreed:

> Everything was possessed of personality, only differing with us in form. Knowledge was inherent in all things. The world was a library and its books were the stones, leaves, grass, brooks, and the birds and animals that shared, alike with us, the storms and blessings of earth.[5]

Non-Indians who have not studied Indians quite commonly think that Indians ascribed spirit or power to nature because they lacked understanding of the real forces at work in nature. Nothing could be more misleading. Indians were not "primitive" in the sense of being less intelligent or more credulous. They had an extensive knowledge, based on experience, of every species of the regional ecosystems they lived in, and an intensive knowledge of how each species interacted with others, including human beings. Though this knowledge was not strictly scientific in the modern sense, it might well be envied by a modern ecologist studying the same ecosystems. The Indians animistic view was in itself a way of understanding and relating to the phenomena experienced in nature, and it often demonstrated intuitive ecological understanding. To take one example, the Indian tendency was to look on processes in nature, even those so fundamental as time itself, as cycles. Change, to Indians, was not an irreversible line, but a curve bending back toward its beginning. This seems to be a radically ecological way of viewing the interactions between human beings and nature, as well as within natural systems themselves. Such ecological concepts as the oxygen cycle, the water cycle, and recycling all follow this trend of thought, and study of Indian ideas may provide further examples. We may still find ways in which American Indian ecological wisdom can be of value in the present day.

American Indians expressed their ecological perspectives in religious terms. The words of some of their holy men show that they thought about this in a sophisticated way; these elders not only honored nature, but also evolved well-integrated explanations of its dynamics. In these explanations, the distinctions between supernatural and natural causes that are characteristic of European-American philosophy are absent.

Their intuitive grasp of ecological concepts was expressed in myths and legends, in a spiritual tradition passed on from generation to generation. The tradition was the means of making the spiritual perception of past experience with nature available in a living way for the guidance of young people who were learning how to act as members of the tribe. The ethics taught by the Indian elders in sacred traditions is an ecological ethics, stressing respect for living things, forbidding needless destruction, and prohibiting waste of the food they have provided. It amounts to ecological good sense based on long experience. Tradition taught restraint, in taking only what was necessary for the environment. It acted as a control on land use, and as a result prevented extinction of species and maintained the natural balance within an optimum range of conditions. The sacred history contained in Indian legends gave an explanation of the world which placed Indians in spiritual kinship with the rest of nature. All American Indian groups used sacred traditions and ceremonial life as vehicles for transmitting and expressing ecological understandings.

Stories, myths, or legends were a major way to communicate the view of nature, as George Clutesi, a Nootka, explained:

> Folklore tales were used widely to teach the young the many wonders of nature; the importance of all living things, no matter how small and insignificant; and particularly to acquaint them with the closeness of man to all animal, bird life and the creatures of the sea. The young were taught through the medium of the tales that there was a place in the sun for all living things.[6]

CHAPTER VII

To the non-Indian mind, these stories are often mysterious. One has a feeling of missing the point, of knowing that there are jokes being told but not seeing what makes them funny, of suspecting that something one would think important has been left out. Reading Indian stories that have been "retold for the modern reader" is one of the best ways of coming up against the chasm that separates two cultures. But within the context of their own societies, these legends explained the world and taught ecology. It is possible for non-Indians to penetrate the cultural barrier by basic humanity, careful study, and sympathetic understanding, and see this.

A Papago "Coyote tale" says that in the days at the dawn of the world the furry trickster was given cactus seeds and told to plant them evenly across the desert. Instead, he flung them carelessly in handfuls only on the south sides of hills.[7] This story points out and explains an observed ecological fact that was familiar to the Papago elders and that their young hearers needed to learn: the distribution of plants in microclimates suited to them. Similarly, the other useful principles of the ethnoscience mentioned earlier in this chapter, the result of close familiarity with animals, plants, weather, and indeed all of the phenomena of nature, could be taught through striking and easily remembered stories.

Many tribes believed that human beings, animals, and plants had emerged from a great cave under the earth, or from a series of lower worlds — the womb(s) of Mother Earth, as it might be expressed, which gave birth to all life. This origin myth not only portrays the oneness and relatedness of all life forms, but the details of the story show careful observation and understanding of the habits and inter-relationships of living things. The order of emergence generally follows the food chain; mountain lions did not come into the world before there were deer. The first animal to emerge from the underworld in one Navajo myth was the cicada, an insect which does come up from underground in the spring. Pueblo legends mention that after the animals had come up onto the earth's surface, some of them made friends, and the coyote and badger teamed up for a hunt. These two animals are in fact sometimes observed hunting together, taking advantage of the mutually useful facts that the coyote can chase swift prey and the badger can dig them out of their hiding places.

A charming Navajo myth symbolizes the return of the seasons in the figure of Changing Woman, an earth goddess engendered by the sun and water, who changes her clothing as the year circles by into dresses of four different colors that indicate the transformations of the vegetation. She grows older, but renews her youth with every coming of spring.

Visions and dreams among the Indians reflected the patterns of their myths and stories, and were widely believed to be ways in which the individual could listen to nature. The natural environment was a source of great power for those who were able to gain it. This could be done by contracting the patron beings, often animals, birds, or fish, who could bestow gifts of power. Elders encouraged the young, especially the young men, to pay attention to these communications, and even to seek them through self-purification and the vision quest. This was expected of all young men, and some mature women as well, in many tribes in several cultural areas, notably the plains and the Northwest coast. We will describe it in these two areas, recognizing that it varied greatly from tribe to tribe and was certainly known elsewhere.

The vision quest was a central element of Plains Indian religion. In this exacting ordeal of self-discipline, a special relationship was sought with the powers of nature, and it was usually granted through an animal or bird that appeared in a vision and spoke special words. The importance of animals in these experiences is understandable, because they played a significant part in the cosmologies of the Plains tribes, as is revealed in the folk stories and legends of sacred history. In some versions, they acted in creation or gave important gifts. In Osage myth, Great Elk made the world on the surface of the water, and the buffalo brought corn to the people. An Omaha explanation of the vision quest is of special interest from an ecological standpoint, since it makes a direct analogy between a man's dependence on animals for spiritual power and human physical dependence on animals in the food chain:

> Man lives on the fruits of the earth; this is true when he feeds on the animals, for all draw their nourishment from mother earth; our bodies are strengthened by animal food and our powers can be strengthened by the animals giving us of their peculiar gifts, for each animal has received from Wakonda some special gift. If a man asks help of Wakonda, Wakonda will send the asker the animal that has the gift that will help the man in his need.[8]

The individual seeking a vision presented himself as one having no power, dependent upon nature and in need of gifts. Before the ordeal, there was a purifying sweat bath, a practice also done in connection

CHAPTER VII

with other ceremonies, and important enough to be recognized as a ceremony in its own right. The vision quest took place for several days alone in a special place chosen carefully for its natural setting, with fasting, thirsting, and offerings of the person's own flesh cut in small pieces from the body. At some time during the retreat, the animal who was to be the individual's guardian spirit would appear and speak, giving a significant message, teaching a song, and designating a special object or design to be used on a shield, a tepee, or painted on the body, that should always be kept as a talisman of the vision and the guardian spirit itself, so that the power could continue to be present with him. Sometimes a name was given, incorporating the name of the animal seen in the vision. Often the visions were very elaborate and seemed to last a long time. Black Elk said, "'Lamenting' (the vision quest) helps us to realize our oneness with all things, to know that all things are our relatives."[9]

The vision songs were not thought to be composed by the individual, rather they were given to him as part of the vision itself. Sometimes the songs were simple, sometimes hauntingly evocative. A Mandan-Hidatsa eagle song contained only these words:

> Above the earth I walk;
> on the earth I walk.[10]

A deer song of the same tribe declared:

> The first snowstorm is good.[11]

And two Pawnee songs ran as follows:

> My brother the fox spoke and said,
> "Behold and see the wideness of the earth;
> The white foxes know the earth is wide."[12]

> I am like a bear
> I hold up my hands waiting for the sun to rise.[13]

A Teton Sioux man had a song of two birds, the crow and the owl:

> At night
> may I roam
> against the winds
> may I roam
> at night

may I roam
when the owl
is hooting
may I roam

At dawn
may I roam
against the winds
may I roam
at dawn
may I roam
when the crow
is calling
may I roam.[14]

An Indian was never the same again after his vision, since he had partaken of the sacred quality of the spirit being that came to help him.

In several tribes, an individual who had had a vision became part of a society composed of members who had received visions of the same animal, and was ever afterward assigned certain responsibilities and bound by the rules of that society. Among the Sioux, the Buffalo Society (*Tatank Watcipi*) was made up of men who had had visions of buffalo. The Bear Society of the Assiniboins may serve as a good example of the type: all its members had received vision power from the bear. They wore stylized costumes with bear claws and two round hair knots that suggested a bear's ears. In war and ceremonials, they made a snorting noise like a bear, "huh, huh." In ceremonials, they shared a feast of bear-beloved berry soup, and enacted a "hunt" of one of the society members who acted like a bear. They painted shields and tepees with bear designs, and carried a sacred "bear knife." Their duties included ceremonials, the cure of certain illnesses for which herbal medicines had been given them in visions, and the making of war and hunting medicine.

Along the Northwest Coast, the vision quest was basically similar to that practiced in the Plains, but displayed some special features that are of considerable interest. There, the protection of a powerful spirit was sought through self discipline, or "training," as many Indians called it, which produced a state of extreme cleanliness, free from the human odors repellent to animals.[15] The suppliant bathed in water and took sweat baths, followed by scraping the skin with rough but sweet-smelling bark, branches and herbs, while reciting formulaic prayers.

CHAPTER VII

Many houses had sweat-baths attached, and a daily immersion in cold water followed by scrubbing was the usual practice, but in the quest for a spirit, these practices were intensified. Complete or partial fasting from food and drink was combined with strong doses of emetics and purges to cleanse the body of any trace of pollution. Both men and women of all tribes observed strict sexual continence at these times; and it was felt that the more it was prolonged, the greater would be the power accumulated. The whole process of purification could go on for days or weeks, even months or years in exceptional cases. The suppliant went out away from the settlement to seek the vision; in the case of the Quillayute, for example, he stood on a high platform built up in the mountains.

Eventually, the guardian spirit would have mercy on the properly prepared seeker and appear, to grant visions and powerful gifts. The spirit could be that of almost any species of animal, bird, fish, plant, geographical feature, or one of the fabulous monsters. The universe as portrayed in myth is three-storied, with places for various beings to live both above the dome of the sky, below the sea, and in the underworld, as well as in the air, land, and sea of the visible world. Each being had its own proper home. Those beings included monsters composed of human and animal parts in striking recombinations; a huge, hairy, man-like cannibal monster among them, and a terrifying wild woman of the woods. Even these would sometimes yield to the seeker and give them special powers. But the most important ones, and those most intimately connected with the daily life of the people, were the powerful spirits of the animals of land and sea. The supplicant was sometimes taken to the home or village of his guardian spirit, where he saw the animals in their human form and perhaps witnessed their dances and the rituals of animal medicine people. Then gifts were granted; power for hunting, warfare, or curing, a name full of prestige for the individual or for his house, a symbolic, heraldic crest representing an animal that could be carved on his house posts or totem pole, a song to sing when he wished to call for power, and many times also a dance to be performed in a mask representing one or more of the forms of the spirit-protector, imitating the actions of the animal that had been revealed in the vision. From that time forth, the person was a relative of at least one of the tribes of nature; of the bears, perhaps, or the killer whales, the raven, sculpin, cedar tree, or the fabulous thunderbird. This did not always mean that he was forbidden to kill the animal associated with his patron; in fact, the vision may have promised that animals of that species would afterwards come willingly and sacrifice themselves for his benefit, as long as he kept up the practices of purification and ritual that were pleasing to the guardian spirit.

Gifts received in visions were treated as hereditary treasures, passed down in family lines from the ancestors who had first received them. Names were valued in this way, and were transmitted with lavish ceremony. The names themselves are as evocative of nature as Homeric similes; "Raven-Cawing-as-He-Flies-Out-to-Sea-in-the-Morning," for example, or "Sunshine-Glinting-on-the-Emerging-Dorsal-Fin-of-the-Rising-Killer-Whale."[16] Clans owned legends of descent from ancestors like the wolf, eagle, raven and killer whale, and used their emblems in their works of art. Reciprocity between human beings and the creatures of the natural world is thus a major theme of Northwest Coast culture.

Young women at puberty were required to live by themselves for a time and practice some of the same self-disciplines of fasting, washing and continence, because they also were receiving a power deeply connected with the cycles of nature, the power of generation, and other powers often became manifest at the same time. The story was told of the Tlingit young woman who looked out of her hut at that time, and was able to call a glacier to come to her, although it was an unwise thing to do.[17]

In tribes everywhere in North America, communication with animal spirits seen in visions and dreams was regarded as the way an individual could gain special power to become a medicine man or woman, called a "shaman" by anthropologists. These people practiced unusual self-discipline and received gifts of surpassing potency from the spirits of nature. They would often perform amazing feats of magic, devise effective tools and traps for hunting, retrieve foreign objects from inside the bodies of sick people, and make spirit journeys in symbolic "canoes" to bring back lost souls. A story often told in many tribes said that the first inhabitants of the earth had been the animals in human form, able to speak and organized into tribes like Indian people. Although they seldom appeared in that guise in later times, a shaman could still see them that way, and address an animal with the attitude contained in the statement, "You and I have the same mind and power."[18] Above all, shamans could summon the power of their animal guardians to wound or to heal, and could use the various properties of many species of plants. They could treat disease, since it was believed to be caused by animals that had been angered by human ac-

CHAPTER VII

tions. In addition to understanding the importance of treating the human mind and body as a single, integrated organism, the Indian medicine men also knew that the individual's relationship to the whole world of nature also controls and assists the healing process.

Dreams were regarded as important experiences by every Indian tribe, and it is almost impossible to distinguish them from visions.[19] Often they are the same; dreams are visions seen in sleep, visions are waking dreams, and sometimes the person who sees one cannot tell which it is. The content is similar. It would be expected that dreams reflect one's unconscious view of the world, and that appears to be the case with Indian dreams that have been recorded. Their themes include animals, hunting and planting scenes, animals that talk and give powerful abilities including leadership, skills as warrior, hunter, or healer, animals that change into human form, visits to the sacred villages of the animals, and virtually every image from nature mentioned in these pages. The dream life of the Indians reveals the numerous sense of oneness with the powerful spiritual world of nature that we have already seen to be the guiding inner principle of their environmental attitudes and practices. It seems that this sense may have been derived from the dreams. Indians prized their dreams and never dismissed them with the contemptuous "that was only a dream" that is heard so often in the dominating non-Indian culture. This might give us pause. Many of the themes of Indian dreams mentioned above also occur commonly in the dreams of non-Indians, but are repressed or rejected as "irrational." It is as if our subconscious minds were presenting us with an experience of the world that is not unlike that of the Indians, and we constantly suppress that experience due to our supposedly civilized cultural conditioning. For although dreams reflect our culture, they are not limited by it, and they do contain images that come from a deeper level that we share with all other human beings and the natural world. Careful attention to our dreams would show us we are not as far from the ancient Indian ecological perceptions as we sometimes think.

The ecological wisdom of Indian tribes was passed on to succeeding generations not only through the stories told by the elders and the visions and dreams that individuals were encouraged to have, but also through the pageantry of group ceremonials. Through songs, costumed dances, and symbolic objects, these ceremonies often enacted the events of a sacred story cycle like the ones described earlier in this chapter. Embodied in the festival of tribes in each of the cultural areas are their knowledge of and their approach to the natural world. Let us look at a few examples.

The very rich series of Plains Indian ceremonials expressed their sense of relationship to the natural environment. Through songs, the smoke of sweet grass and tobacco, dances, purifications, and the offering of gifts, especially those afforded by their own physical and spiritual lives, Indians believed they participated in nature and contributed to its periodic renewal. Ceremonies included acts of honor to the six sacred directions, west, north, east, south, up, and down, in clock-wise order because that is the direction the sun moves through the sky. Sacred bundles contained parts of animals and plants embodying the power in nature; designs of animals, plants, and other natural objects decorated ceremonial regalia, tepees, shields, and clothing and other things made for daily use. These designs not only recalled the inhabitants of the natural world, but were believed to help make their power present.

The sacred pipe of the Sioux also expressed their attitude of dependence upon the natural environment. According to their traditional history, it was given to the people by a woman who embodied the spirit power of a white buffalo calf, so that Wakantanka gave his great gift through an animal. The pipe itself was a complex symbol, representing the universe and human beings at the same time. The bowl was round like Mother Earth (the horizon where earth meets sky); a buffalo calf often carved upon the stem stood for all animals, and tobacco flakes represented all birds. At the same time, every part of the pipe was held to correspond, microcosm and macrocosm coincided. The Mandans had a "women's band of the white buffalo cow,"[20] whose dance was held to attract the wild herds near the village.

The most widespread and characteristic communal ceremonial of the Plains Indians, the Sun Dance, is a way in which the people express their participation in the great cycles of nature. "Sun Dance" as a name expresses only a small part of its purpose; it is really a world renewal rite associated with the appearance of new green vegetation and the increase of animals, particularly the buffalo. The Cheyennes call it "New Life Lodge." It is a quest for power and a renewal of communion with the earth, done by many people at once for the whole tribe. In fact, it is the vision quest in the form of a social ceremony rather than an individual ordeal. The flesh offerings sometimes given are not required, even though they are the best-known part of the dance to outsiders. Among some tribes, the flesh is not pierced. The discipline of dancing

CHAPTER VII

for several days and nights, and abstention from food and drink, are held to be self-sacrifice enough to demonstrate sincerity and need to the powers of nature. Each dancer holds to a path toward the center pole, blowing on an eagle-bone whistle that calls upon the Thunderbird for rain and new vegetation. Visions are awaited patiently and earnestly. The center of the Sun dance is a forked cottonwood tree, chosen and cut with rituals and addresses of honor, set upright in the designated place and decorated with green plants and the skull of a buffalo. The pole symbolically connects earth and sky. The four sacred directions are also indicated; the lodge opens toward the rising sun, so that the lodge as a whole represents the natural universe in whose renewal the people are participating. The Sun Dance, along with many other rites of the Plains Indians was intended to propitiate the buffalo, which was at the same time the main support of their lives and the most pervasive spiritual force they perceived in their environment.

Among the Pueblos of the Southwest, ceremonials followed the cycle of the agricultural year, and were directed toward maintaining the balance of nature, producing rain, and aiding the growth of crops. The Papagos say they dance to "keep the world in order" and prevent floods.[21] Pueblos have complex calendars of ceremonial dances and ritual enactments whose dates are set by the sun's movements. With songs, masks and costumes, highly disciplined group dances, offerings of prayer sticks, food, feathers and corn meal, and secret rituals in the kiva chambers, including prayers and ritual speeches, the Pueblo Indians seek to unite and concentrate the good desires of the people for rain and growth. If properly done, they firmly believe, the ceremonies will achieve their purposes. They could only fail through mistakes in repeating the ritual, or evil thoughts that intend to hurt someone else. Pueblo ceremonies have more than their solemn side, however; clowns are present, too, and given freedom to flaunt humorously every convention of ordinary Pueblo society. Many of their skits and practical jokes have a sexual or scatological reference (which should perhaps no longer shock Anglo Americans), not unconnected with the ideas of rain and fertility. Pueblo rituals are not orgies in any sense of the word, but beautiful stylized dramas that include tragic, comic and liturgical elements. A good ceremony, the Hopis say, is "for the benefit of the whole world,"[22] and one participates simply by being present with a good heart.

Almost all Pueblos have dances of masked, costumed kachina figures. These may represent any spiritual beings known to the Pueblo

Indians, including gods, animal spirits, clouds, sun, and spirits of plants, birds, insects; indeed, almost any recognizable entity of the natural environment. Some kachinas exist as spiritual beings in their own right. Several hundred different kinds have been catalogued. Their masks are highly symbolic and stylized, designed to represent the essence of the spiritual being depicted, not just the outward appearance of the animal or other creature. Following is an example of a Hopi kachina song, typical in its expression of assurance that a response from the natural environment will come:

> The green prayer-stick brings the water
> For the earth and its vegetation are combined in it.
> From the four corners come the clouds —
> Come together, gather over us.
>
> The green prayer-sticks bring the water
> From the four directions in which we planted them.
> The Spirit of the Rain passes over the prayer-sticks,
> And their feathers are stirred.
>
> We have found the water,
> It has entered to the roots,
> All things are beautiful,
> All things are glad.
>
> Good now! Here am I, Father Sun, with my prayer-stick,
> We are asking you for drink.
> The glisten of running water is beautiful;
> Let the quickening rain, the heavy rain, come.
>
> We have found the water,
> All vegetation is beautiful.
> The water has entered the roots,
> The Spirit of the Crops is happy.
>
> The Father Sun is watching us;
> With his rays comes the water;
> The green prayer-stick has brought it.
> The corn is beautiful. We are glad.[23]

When they witness Pueblo dances, non-Indians often assume that they are done simply to bring rain and encourage the growth of the crops of one village. But when the Pueblo elders themselves were asked to explain the purpose of the rituals, they replied that they did them for the whole world, for a good relatonship of all human beings to all of nature. Just as non-Indians who attend Pueblo ceremonies "with a

CHAPTER VII

good heart" are believed to help them be effective, so the Pueblos believe they are helping all vegetation to grow, even the gardens of non-Indians.

Nowhere were tribal festivals more spectacular than among the Northwest Coast Indians. The living entities of nature that provided them with both sustenance and power also formed the themes of their art: drama, dance, music, sculpture in wood and stone, painting, and weaving. Ceremonial life in this area was really theatre, including masked and costumed figures, dancing, songs, dialogue, and visual effects. It took place only in the winter, since the powerful beings represented were believed to enact an annual cycle of travel, appearing in the villages only during that season when the activities of food procurement and preservation reached a temporary low point. The ceremonial theatre presentations were given by dancing societies, each of which owned one or more of the complex rituals, with the costumes and other paraphernalia required. Dancers wore masks and costumes that identified them with their animal guardian spirits. For example, Haida and Kwakiutl dancers impersonated the grizzly bear, moving in rhythmic imitation of a bear's swaying stance and gait. A Nootka Wolf Dance re-enacted an ancestral visit to the legendary House of Wolves and the granting of ancient prerogatives there. The ceremonies usually involved the ritual "capture" of young novices and their initiation into the society concerned. The "Dionysiac" theatrical cannibalism that was feigned in some of these ceremonies should be understood in light of the close animal-human interchangeability that permeated their thoughts and feelings: killing and eating animals was absolutely necessary to life, but it inevitably involved the idea of cannibalism (so, indeed, does Christian communion, a fact that led some missionaries to refuse to introduce it to their converts on the Northwest Coast, fearing that the Indians would confuse it with their old practices). Actual cannibalism was subject to severe taboo, and was unknown outside the theatrically convincing, but purely symbolic act in the ceremonials.

Dancers representing animals were present in every part of the Northwest Coast culture area. The Yurok of northern California, as part of their cycle of ceremonies intended to renew the world and assure its abundance, held a White Deerskin Dance, in which the deerskins were carried on poles and moved in a way incredibly evocative of the living deer.[24]

Music on the Northwest Coast was both instrumental and vocal. Plank and box drums, tambourines, rattles, whistles, horns, and the bull-roarer were used in part to imitate the calls of animals and birds, and other sounds of nature. Songs were sung individually and chorally; early explorers remarked upon the beauty of songs drifting across the water as canoe loads of Northwest Coast Indians approached, not a common reaction when White ears first heard non-Western song in most parts of the world. Many of the song texts were framed as if the animals themselves were singing them, as, for example, the Nootka Song of the Hair Seals: "When we come to a big rock, we all sing together," or the Bear: "I am hungry for salmonberries," or the Eagle: "I am dancing in the air and dancing round and round." Others expressed the awe of the beholder of the shark: "Where are you, on whose back the waves break," or the whale: "Darkness as of approaching night on the water."[25]

The highly conventionalized representation of animal forms in the sculpture of the northern region, with its ability to transform human shapes into animal shapes with a slight change of perspective, and interlock them with each other, is perhaps the most widely known aspect of Northwest Coast art. This art was a spectacular series of semi-abstract, highly developed displays, demonstrating their perceptions of the natural world as seen in the visions they had received. On carved poles, animals and birds are distinguished not by their naturalistic resemblance to living creatures, but by their portrayal with one or more distinct features from a "vocabulary of attributes." For example, the Raven has a straight beak, the Eagle a curved beak, and the Thunderbird a curved beak and horns on the top of the head. The Hawk has a strongly recurved beak and an extra, "human" mouth. The Beaver has two long, squarish incisor teeth, a diagonally cross-hatched flat tail, and often a cylindrical stick held crosswise in the front paws.

Equally intriguing, though less well known, is the subtlety and sophistication of painting on flat surfaces such as boxes and house screens. Here the strong bilateral symmetry of sculpture in the round is maintained, but there is even less literalism. All the "necessary" parts of the animal, internal and external, are represented, distributed across all the available space, and the "unnecessary" parts left out. Both sides of the animal are shown, so that the design may appear to be two animals facing each other, if indeed the uninstructed observer can see an animal at all in the elaborated design. Each part is represented in a conventional way; a joint, for example, may be shown as an eye.

It should be evident that these works of art are not attempts to make pictures of animals, but ways of portraying the animals' essential

CHAPTER VII

power. They are carved or painted because they are crests belonging to the person for whom the objects are made. They may be his clan animals, his guardian spirits, or they may take part in one of the traditional stories that are his by right. They are constant reminders to the Northwest Coast Indians that all living things are sacred beings, the ultimate bestowers of all gifts, physical and spiritual.

Art in every cultural area had the same function. The colorful Kachina dolls of the Pueblos, the painted figures that adorned the tepees on the plains, and the giant forms of humans and animals limned by Colorado River Indians on the surface of the desert all served as visual aids in the communication of environmental consciousness, one of the ways in which the wisdom of the elders appeared anew in the succeeding cycles of years.

Ceremonials of Nature. The Shoshone Sun Dance is held in a lodge that is built especially for this ceremony. The lodge symbolizes the universal environment. Photographer A.P. Porter, 1910.
COURTESY SMITHSONIAN INSTITUTION, NATIONAL ANTHROPOLOGICAL ARCHIVES.

Ceremonials of nature. Arapaho Indians pray during the Ghost Dance. Taken in Oklahoma territory by photographer James Mooney, circa 1890.
SMITHSONIAN INSTITUTION, NATIONAL ANTHROPOLOGICAL ARCHIVES.

Designs from nature. A Tlingit man, named Sitka Jake, wears a Chilkat robe and dance headdress, and carries a rattle in his hand; the decorations of which represent natural forms of life.

COURTESY SMITHSONIAN INSTITUTION, NATIONAL ANTHROPOLOGICAL ARCHIVES.

Ceremonials of Nature. A buffalo skull rests at the base of the central tree-trunk pole used in the Sun Dance. This ceremony celebrates the rebirth of nature, using many ritual objects taken from the natural environment. Photographer Donald A. Cadzow, 1926, at Piegan Reserve, Alberta, Canada.

COURTESY MUSEUM OF THE AMERICAN INDIAN, HEYE FOUNDATION.

Ceremonials of Nature. A sun dance of the Teton Lakota Indian tribe blows rhythmically through an eagle-bone whistle. Photographer Bert Bell, 1929, at Rosebud Agency, South Dakota.

COURTESY SMITHSONIAN INSTITUTION, ANTHROPOLOGICAL ARCHIVES.

VIII

Our People Covered the Land

AMERICAN INDIANS felt that their population was sufficient for the land, in fact they often expressed the idea that they were exactly of the right number, not too many, not too few. As Chief Seattle put it, "Our people covered the land as the waves of a wind-ruffled sea cover its shell-paved floor."[1] The Pueblo Indians believed that every birth will be balanced by a death in the endless cycle of human lives, otherwise there would not be food enough for everyone.[2] The Indians apparently recognized the desirability of keeping their population in balance with what modern ecologists would call the "carrying capacity" of the land. A Cherokee legend represents the animals as worrying that people were becoming too numerous, and burdening Mother Earth.[3] So the possibility of overpopulation did occur in the Indian world view, even if it was rarely a problem. Even in the presence of natural abundance, there seems to have been no social pressure for the production of offspring beyond the minimum number needed to secure the continuation of family and clan lineages. So Indians kept their population low enough so that generally speaking, pressure on natural biotic systems was well within their capacity to bear.

But Indian numbers were not as low as some estimates of early demographers indicated. It used to be repeated in anthropological and historical texts that the total population of Indians in all North America above the Rio Grande in pre-Columbian times was about one million. This low figure suited the ideas of some individuals who were so bound up within the mind set of modern Western civilization that they believed it is human nature to expoit the earth, and could not fathom the evidence that shows that Indians did not damage the natural environment because Indians acted in accord with their traditions that regarded nature as filled with spiritual powers. When faced with this evidence, these people often maintain that the Indians' failure to damage the environment was the result of their small population

CHAPTER VIII

and lack of technology. They say that if the Indians had been more numerous, or had possessed more efficient machines, they would have been as destructive as the Europeans and their cultural descendants have been. Let us look at the first idea: the belief that Indian populations were miniscule.

The method followed by the early demographers who estimated Indian populations was to seek the lowest verifiable figures. Their method was to establish a dependable minimum. This was a valid, defensible scientific method, and it was not misleading as long as those who used it remembered that it arrived at exactly what it tried to establish: an irreduceable minimum, not a reasonable estimate of the total number of Indians before Columbus. Scholars who have been studying the problem of New World populations in the last few years have come to maintain that much higher estimates must now be made than formerly.[4] A major reason for this is that epidemiological studies have recently been made that show the magnitude of the effects of European diseases on Indian populations after the time of first contact, but before the wave of colonization. In addition to the diseases that were deadly to Europeans, such as smallpox and tuberculosis, there were "common childhood diseases" like measles that usually failed to kill Europeans, but to which Indians had developed no resistance. These literally decimated whole villages and tribes, often when there were no Europeans around to see and record the scenes of horror. Diseases spread from one tribe to another, so that Indians who had never seen a European died of European diseases. Archaeological studies of Indian burials now indicate a large aboriginal population, compared to the minimum figure given above. Estimates still vary, but a reasonable figure seems to be well over five milion and probably close to ten million. California alone probably had one million inhabitants. Low population estimates have always served the purposes of those who would downgrade the importance of Indians and imply that the non-Indian settlers occupied a land that was practically empty.

But even ten million people was well within the number that could live in balance with the bountiful ecosystems of North America. And the fact that Indian populations did not increase to the limit of the ability of the land to sustain them is due, to a great extent, to traditional practices and conscious choice rather than starvation, which was rare, or warfare, which was notably less lethal than the European type, or to the rate of infant mortality, which was unfortunately high by modern standards all over the world at that time.

One tradition that doubtless acted as a means of population control was sexual abstinence. This was practiced widely in preparation for and during hunting, warfare, and religious ceremonies. This was done not because sex was thought to be bad; exactly the opposite was true. Sexual energy was deeply respected as a form of personal spiritual power that ought to be conserved, and used only sparingly so that it might not be used up or dissipated. Sexual continence was practiced for long periods of time to accumulate power and good luck, which came not as a reward for "going without," but as a natural result of conserving one's energies.

Most Pueblo ceremonies required the participants to observe sexual continence for a period of several days, and there were very many ceremonies. In the Northwest, the owner of a salmon weir would remain continent during all the time of its construction. Some men kept away from their wives for an entire year, since accumulation of wealth was believed to be proportionate to the length of the period of abstinence. But continence was a necessity for both husband and wife, and a man's abstinence would be invalidated if his wife were unfaithful. Indeed, a disaster would follow.

This is an illustration of the generally strict sexual morality that was observed in Indian families. Seduction of wives was never taken lightly, and marital chastity was highly prized. Infidelity was severely punished, so that adultery was rare. Even here, however, the motive was to preserve inner power rather than to avoid guilt.

The number of children that an Indian wife bore and raised to maturity was small, as a rule, and the Cheyennes provide an example of the fact that this was often deliberate choice on the part of the parents:

> The man of strong character and good family vows at the birth of his first child (especially if it is a boy) not to have another child for either seven or fourteen years. All of the father's growth powers are then concentrated on the development of this one child rather than being dissipated among several. It is necessary to understand that more than the semen of conception goes into growing a child; there is a continuing transfer of the father's "energy" from parent to offspring. . . . Should a parent break the vow of dedication, it is believed it will kill the child.[5]

Perhaps not many couples assumed such a strict vow, but the practice is one among a great number that tended to restrict population increase.

CHAPTER VIII

All tribes had methods of avoiding conception, including an excellent knowledge of the menstrual cycle and medicines, usually plants.[6] A few of the latter have been shown to be effective in modern laboratory tests with animals. Most have not been tested at all, but since Indian medicine has given us many useful drugs, from ephedrine to quinine, it would seem wise to investigate more of them.

The second popular reason given to explain why the early Indians did not severely damage the environment is their lack of technology. Before Columbus, they lacked the plow, any beasts of burden larger than the dog (and, in South America, the llama). The wheel sometimes appeared on toys, but never as a useful machine. Needless to say, they had no guns or explosives. So their critics say the Indians were simply incapable of harming nature very much, and that it was this inability, rather than their religious perspective and conscious decisions, that kept the continent unspoiled. But this argument ignores the technology the Indians did have. The bow and arrow constitutes an efficient killing device. Even when the Plains Indians had acquired the rifle, they continued to use the bow and arrow to shoot buffalo from horseback because it could be "reloaded" more quickly than the rifles then in use. And the Indian hunter was a formidable expert.

Indian technology was certainly capable of doing more damage to the environment than was actually done. Such skilled hunters could have depleted the birds and animals. Fishing weirs could have been left closed, instead of being opened to let many of the migrating fish through. It has been demonstrated that a stone axe in the hands of a man who is used to it can cut down a tree as fast as or faster than a steel ax. It was not lack of technology, but the basically ecological attitudes of the Indians, that prevented exploitation. This made Indians careful hunters, even armed with rifles. In the words of Roderick Nash, "Even with a power saw in his hand there is reason to suspect that the pre-Columbian Indian would have been environmentally responsible."[7]

But what about the cases in which the population was thickly clustered in small areas: the villages, towns, and cities? Here the human impact on the environment was concentrated, so we can look either for damage or methods used to prevent it. We find both.

Iroquois villages stood in forest clearings, protected by palisades made of saplings. Within these protective walls stood several bark-covered longhouses, each with several fireplaces and space for several probably related families. Longhouses are well-named. They were typically 25 feet wide by 50 to 100 feet long, and the foundations of one

400 feet long have been excavated. Crops were grown in the cleared area outside the palisade, and hunting and fishing took place in the forest surroundings. After a while, such a large village probably threatened to deplete the local environment; the soil may have lost its ability to support vigorous agriculture, the nearby timber they preferred for construction and firewood may have been used up. For these reasons, and perhaps also for health considerations, the villages were moved from time to time over relatively short distances of a very few miles. Archaeological studies show that this happened on an average of once a generation (about once every 30 years).[8] The forest was allowed to reclaim the place, and the resources of soil and timber renewed themselves. Thus an Iroquois tribe like the Onondagas could "recycle" the site of villages within a well-known territory over a very long period of time.

Much larger concentrations of population lived in the great centers of the mound-building civilizations in the Mississippi Valley between 500 and 1500 A.D. Their cities contained numerous pyramid-like earth structures that resemble others built of earth and stone in Mexico and Central America. Like them, they supported temples and/or contained burials. One can imagine the numbers of workers needed to raise something like Monk's Mound in Cahokia, a prehistoric settlement in Illinois. This is a single pyramid, with a base measuring 700 by 1000 feet, a height of 100 feet, and a volume of 22 million cubic feet. Only two other pyramids in North America, both in Mexico, exceed its volume. Cahokia contains 120 mounds, many quite large, and it is surrounded by other major and minor population centers of the same period. These people lived in a rich agricultural area and must have used it to the full to produce the crops needed for a considerable population. They had "a complex society that for centuries controlled the natural resources of an immediate hinterland"[9] and had a far-reaching trade network that brought, in the case of Cahokia, flint from the Ozarks, copper from Lake Superior, mica from North Carolina, and shells from the Gulf of Mexico. But by the time the European explorers arrived, only one tribe, the Natchez, still lived in towns with mounds and temples like those of earlier times. What caused the disappearance of the Mound Builder civilizations could have been an ecological disaster, according to some modern students of the problem. Exhaustion of resources such as timber for the buildings, deer and other animals in the parts of the forest nearest to the population centers, and soil exhaustion in the sections used continuously for agriculture, have been suggested. While these events seem unlikely in the light of the ecological attitudes and practices we have seen everywhere in this book, we must not forget that elsewhere in the world the rise of urban societies and the greater concentrations of people they create have corresponded to the multiplication of human impacts on the natural environment. "Population concentrations and the overuse of local resources go hand in hand."[10]

Other than the Mound Builders, Pueblo Indians had the closest approximation to urban life north of Mexico. Comprised of stone structures often several stories high, pueblo dwellings have been compared to apartment buildings. But the Pueblos stayed close to the land, and the distinction between urban and rural was never made.

Pueblo architecture is uniquely suited to the Southwestern environment, using readily available materials. Spanish and Anglos in New Mexico have widely adapted it, and with good reason. The natural insulation of thick stone and adobe walls made pueblo rooms cool in summer and warm in winter. Pueblo buildings blend harmoniously into the landscape; from a moderate distance, it is hard to distingush Hopi pueblos from the rocky outcroppings on which, among which, and out of which they are constructed. This is reflected in Hopi terminology; the word for "inner room" means "cave" and the word for "terrace" means "ledge."[11] To the Hopis, building a house is not an act of conquest over nature; it is an act of reverence toward nature, accompanied by rituals. To establish the outline of the house on the ground, pieces of cactus are buried in the corners, along with prayer-sticks, to give the house "roots."[12]

Pueblo town planning uses the plaza, an "open kiva," a theater for sacred dance, as its center. Prayer-sticks are also buried there as the "roots of the town." Buildings cluster around the plaza or form rough lines on either side. The orientation of houses and plazas may well have been planned in relationship to their views of the impressive sacred mountains on the horizon.

Sanitation in the early pueblos has been criticized, but it was probably no worse than in European towns of the same period. In some respects it was worse, with burials under the earthen floors and abandoned inner rooms used as trash receptacles. In other respects it was better. The dry climate and clear, strong sun of the Southwest helped. Although trash was thrown down the cliffs, almost no food was wasted, and all waste material was biodegradable or, like the ubiquitous potsherds, "earth returning to earth." The most durable and dangerous components of modern urban waste were missing. A pueblo trash heap soon assumed its familiar appearance: a mound of earth covered by trees and plants. Pueblo Indians, unlike their European contemporaries but like the Southwestern Indian neighbors, were scrupulously clean. Their ritual observances demanded frequent baths, sweat baths, and hair washing with copious suds from yucca roots.

Water supply for most pueblos was carried from clear, unpolluted streams, or springs that received periodic ritual cleansings. Water pollution was strictly avoided. The only air pollution known was wood smoke; centuries of experience had taught the Pueblo Indians how to design kivas and other rooms with shafts and deflectors to provide fresh air and cause the smoke to rise out of the smokehole-doorway in the flat roof.

Population increase does not seem to have presented a problem to the Pueblo Indians. Their towns stood for centuries in the same locations, and some still do. If a town really did have too many people for the cultivable land in the vicinity, it could divide or, less probably, move to a new spot. Men of Oraibi are known to have grown crops at Moenkopi, forty miles away, in historic times, and eventually founded an additional town there.

As with the Mound Builders, though, archaeologists have tried to explain why the ancient people of the more northerly pueblos abandoned their homes at Mesa Verde, Chaco Canyon, Betatakin, and elsewhere during the late thirteenth century. Changes in the natural environment may explain this widespread and important movement. Other causes, such as the actions of aggressive enemy tribes, have also been suggested.

Climatic changes in the Southwest certainly have occurred.[13] Studies of pollen deposits show that juniper now grows in places where spruce once was common, and geology reveals that there have been periods of erosion and periods of alluvial deposition.[14] Tree ring studies demonstrate periods of relatively abundant rainfall and periods of drought. A detailed chronology of Southwestern climate remains to be worked out, but there seems to be general agreement that the latter half of the thirteenth century was an exceptionally dry period. A Great Drought (1276-1299) has been postulated from tree ring studies. Floodwater farmers probably experienced many crop failures through lack of rain, and the general reduction of vegetation may have triggered a period of accelerated erosion, beyond the ability of Pueblo methods of control.[15] Erosive cutting of deep arroyos would have made water inaccessible to many fields, as violent summer thunderstorms continued to bring flash floods at intervals even in years of drought.

The most recent erosion cycle, which began around 1875 and has again produced deep arroyos, is at least partially blamed on overgrazing by domestic sheep, goats, cattle and horses.[16] This was, of course, not the cause in the 1200s, but cutting of trees for firewood and construction in the pueblos has been suggested as a cause of deforestation and resultant erosion, and as a contributing factor in the abandonment of the northern pueblos.[17] This would mean that Pueblo Indians contributed to their own decline through their impact on the natural environment. Pueblos, like other Southwestern Indians, had deep reverence for trees, and would in any case prefer dry driftwood and dead timber for firewood. Living piñon pines (a food source) and junipers are found even today within easy walking distance of the Hopi towns. Pueblo depletion of a flourishing mixed forest like that of Mesa Verde seems very unlikely. Ponderosa pines once grew near Chaco Canyon, and the Pueblo inhabitants used hundreds of them in constructing their buildings, but they did so over a period of 200 years or so, adequate time to allow forest regeneration if the climate had not been changing. But the climate did change. Today there are no ponderosa pines in the Chaco area because it is too dry. Studies at Grand Canyon have shown that at the lower margin of their range, small differences in local availability of water are critical in determining the place where ponderosa pines will grow, so that dessication of climate would seem to be an effective factor in changing the ground cover in such locations even when the decrease in rainfall is relatively slight.[18] With drying climate, removal of even a few trees might well have had a serious effect that could not have been foreseen. An interesting modern parallel occurred in 1906, when greatly accelerated erosion of Oraibi wash began as a result of failure to apply the usual methods of conservation: White pressure had caused internal dissension that divided the town, and cooperative labor projects ceased for a time; the floods raged without check, and soon the arroyo was too deep to be reclaimed.[19] It is possible that ancient erosion-control works may actually have delayed the abandonment of some northern pueblos, and that some other factor caused the people to abandon their check-dams and terraces for long enough to allow erosion to advance beyond any chance of restoration.

Hostile enemies are a possible cause contributing to the withdrawal. The period of Navajo and Apache arrival in the Southwest occurred at about the same time; the Utes had always been neighbors. During the

Spanish period, these tribes visited the pueblos peacefully, trading the results of their hunting and gathering for agricultural produces as long as times were good, but at other times they were raiders. During a drought, both cultivated and wild foods would have been depleted. There would have been less reason for trade, and more incentive for raiding. The ecological balance between the two groups thus would have been upset.[20] The theory of abandonment through enemy pressure, however, does not fit the absence of much evidence of violence or looting in the abandoned pueblos; the people left in an orderly fashion, taking most of their possessions. They are believed to have gone to the desert and river pueblos in the south, few of which were in locations as secure as those they had fled. Most Hopi towns were on the flat lands beneath the cliffs until the arrival of the Spaniards caused a defensive movement up to the mesa tops. Still, the northern cliff dwellings have an undeniable "fortified" look. Vincent Scully has recently suggested that their architecture was dictated by ceremonial rather than defensive considerations.[21] As dry conditions set in, he believes, the people moved down into the canyons, closer to Mother Earth, in locations that seemed to resonate better with the natural environment. When this did not improve conditions, they left for the pueblos in the south that were better supplied with water.

Drought, rather than environmental depletion by concentrated populations, seems to be the best explanation for the abandonment of the northern Pueblo areas. The pressure of drought may have meant that methods of coping with the semi-desert environment that worked well in a slightly moister climatic period no longer functioned when drier times arrived.

To see American Indians in truly large cities and empires of the pre-Columbian past, one must travel south of the Rio Grande and therefore outside of the scope of this book. The fascinating story of the Olmecs and Mayas, Zapotecs and Mixtecs, Toltecs and Aztecs, cannot be told here. Especially worth study at another time would be the Inca Empire of Peru, which seems to have provided for the food, clothing, and shelter of several million people in a way that prevented starvation and yet did minimal damage to the ecosystems within which they lived. The carefully constructed and maintained terraces of Inca agriculture kept erosion from occurring, and produced an annual food supply perhaps twice as great as that provided by Peruvian plant agriculture today. The Inca government even included a branch that corresponds to a "Department of Conservation." Study of ecology in the Inca

CHAPTER VIII

civilization might well reveal one case in which a strong central government tried to inform its enactments with the environmental ethics of the American Indians.

An Arapaho Indian tepee was fashioned of buffalo hides (with the hair inside) and shaped to stand up to the strong winds of the plains. Also shown: buffalo meat drying in the sun. Photographer William S. Soule.
COURTESY SMITHSONIAN INSTITUTION, NATIONAL ANTHROPOLOGICAL ARCHIVES.

In the Kwaliutl Indian village of Alert Bay, British Columbia, the boats, docks, houses and poles are all made of wood from the cedar trees from the surrounding forest. Photographer N.B. Miller, circa 1888.
COURTESY SMITHSONIAN INSTITUTION, NATIONAL ANTHROPOLOGICAL ARCHIVES.

Walpi, a Hopi pueblo in Arizona, is constructed of the same stone as the outcropping on which it stands, thus blending into the natural environment. Photographer William H. Jackson, circa 1899.

COURTESY THE LIBRARY, STATE HISTORICAL SOCIETY OF COLORADO.

The Powerful Animals. In the Animal Room at Zuñi pueblo, New Mexico, animals and birds, which are referred to as powerful spirit beings, are represented in the wall paintings. Photographer Adam Clark Vroman, 1900.   COURTESY SMITHSONIAN INSTITUTION, NATIONAL ANTHROPOLOGICAL ARCHIVES.

IX

The Strangers' Ways

TWO COMPLETELY different ways of seeing the earth and treating the earth have existed side by side in America for the past few centuries. American Indians and European-Americans lived in the same land, but neither clearly understood the attitudes of the others toward that land, or what the others thought they were doing with it and with the forms of life and natural resources on it. As the great Nez Percé, Chief Joseph, said, "We were contented to let things remain as the Great Spirit made them. They were not and would change the rivers and mountains if they did not suit them."[1] The two peoples really lived in different worlds, since human beings experience the world in terms of their own perceptions and points of view. Though each spoke to the other, neither realized how different the ideas were that underlay the words used by the other.

The people who came across the Atlantic carried with them a traditional European view of the natural environment, which denied that there is any spirit in nature, or anything intrinsically valuable about the things in nature — and they did refer to natural objects and life forms as "things," never "persons"' or "beings." Therefore they felt that they could use any creature in the natural environment without concern for the creature itself or its role in natural systems. Only its utility to human beings, especially, or perhaps only, to the individual owner of the land where it was found, was considered. This attitude is still dominant today. Non-Indians often assume that Indian conservation practices must be a result of economic necessity or expediency, because non-Indians characteristically think in those terms themselves. They do not share the traditional Indian view of the world as a place where spirit power, ritual, and reciprocity with animals and plants are operative.

Land use became the sharpest point of conflict between the two perceptions of nature. The Indians had gained their sustenance

CHAPTER IX

through a living interrelationship with nature. They loved their land, respected the birds and beasts, and celebrated their relationships in traditional lore and a round of ceremonies, as the previous chapters have explained. When they were told they would have to give up what had sustained them, they replied in words like these: "The Master of Life has given us lands for the support of our men, women, and children. He has given us fish, deer, buffalo, and every kind of birds and animals for our use . . . but certainly it was never intended that we should sell it or any part thereof which gives us wood, grass, and everything."[2]

European-Americans had a different view of what land was for. It was to be held in exclusive private ownership. It was to be bought and sold. It was to farm. It was to settle down on and use just as the owner saw fit. They came into a region inhabited by people who had a completely different view of land and everything around them: the wildlife, the air and the water. They never really understood how different the Indians' view was, or that that view could have any validity or importance, because they saw Indian cultures as essentially flawed and inferior. As part of the intellectual baggage brought over from Europe were concepts such as "barbarian," "savage," and "pagan," which they applied to the Indians. In their view, Indians, though undoubtedly human, did not really count as "human beings" in the full sense. Indian religions were regarded as superstitious; Indians were "lesser breeds without the law," but not without things of value, particularly the land and its resources. The desire to take and use Indian resources is characteristic of many European-American actions in regard to Indians in every period of American history, and the land issue has always been in the forefront of Indian-White relations.

The complete complex web of human relationships to the natural environment that constitutes the cultural ecology of a people is destroyed when the land is destroyed, or when people are deprived of the use of the land. Because of their feeling of gratitude and care for the land, Indians generally were shocked when they first saw the strangers slicing into the ground and turning it over with plows. The breast of Mother Earth should not be torn so violently, they told their new neighbors, but carefully caressed with a digging stick or hoe. Decades of soil erosion, brown rivers, and dust bowls that followed the plow seem to show that the Indians had good reason for feeling the way they did.

Plowing was not the only White activity that outraged the Indian sense of the sacredness of nature and the careful treatment of all beings that they believed was necessary for mankind. Among the first wave of White men to enter the interior forests were fur trappers intent upon killing as many fur-bearing animals as possible and selling their pelts. Of course, the trappers would rather have had the Indians work for them than have to undertake all the arduous labor themselves, so they began to trade with the Indians and apply economic pressure to try to get them to join in their extractive raid on animal resources. If nothing else had upset their way of life, most Indians would have refused. But they were subject to an almost total onslaught by all aspects of White European-American culture. They caught unknown diseases which were initially beyond the ability of the medicine men to cure, their beliefs were ridiculed and undermined, missionaries taught them the error of their ways, and the military and legal might of other people was extended over them. Their very survival is a wonder of history. So the steady diminution of the Indian land base, and the depletion of what remained in Indian hands, acted to fragment and destroy the enrironmental attitudes and practices described in the previous chapters. A way of life came close to disappearing, though it has survived in some important aspects. As soon as their ways of life were disrupted and their ecosystems fragmented, the ecological harmony they had experienced before became impossible. The immeasurable loss to the Indians is clear; what has not always been so apparent is that White people have also suffered loss in the process. "The White man's failure to establish a close relationship with the native civilization is reflected in his tragic disregard for the integrity and beauty of the land he has conquered."[3]

A few non-Indians did come to appreciate the Indian oneness with the "sacred circle" of the natural world around them. As Christopher Vecsey observes, "Even whites taken captive by Indians often found Indian 'nature' ways alluring and refused to return to white society. To them and to the thousands of whites who ran away to live with and like Indians, Indian life represented a return to nature from which white civilization was alienated."[4]

Indians looked at history less in terms of time than in terms of place and people. They saw the past as giving them roots in a particular land, with its air, water, weather, and living things, all connected with the events that marked their life as a people. To them, the life their ancestors lived in their own land before the strangers came was still close to them. As long as they kept their relationship to that land, they

CHAPTER IX

had a living continuity with their own past. As they would say, it was "to us" that the things about to be described were done.[5]

In our discussion of Indian environmental attitudes and practices, an attempt has been made to show that conservation was part of aboriginal American ways of life. But Indian cultures were not static, or suspended in the never-never land of the "ethnographic present." Before contact became an overwhelming onslaught, Indian tribes incorporated new traits and ideas into their own cultural framework without destroying it. Not all foreign ideas adopted by Indians are a corruption of their own traditions. Adaptation and borrowing do not necessarily destroy the cultural integrity of a people; if they did, there would be no valid culture left anywhere on earth, because all contain borrowed elements. This is true both of material and non-material culture. But while borrowed elements may have become authentic parts of the Indian world view, Indian thought had its own complex development often quite alien to European thought. Not every idea that sounds sophisticated or surprisingly "civilized" in an Indian setting is necessarily borrowed from others. As an example, it was once customary for anthropologists to dismiss every tendency toward a monotheistic world view among Indians as the result of exposure to Christian missionary teaching. A better approach would carefully examine the evidence for each case separately. Even when an idea was borrowed, it often took on a different function in the new context.

Contact had great influence on agriculture, with new plants like wheat, chili, and fruit trees, and new domestic animals, spreading rapidly from tribe to tribe even before they met the Europeans. The Pimas became prosperous wheatgrowers in early times, and the Havasupai were already growing peaches when the first Spanish missionary visited them in their alcove of the Grand Canyon. In some places where the plow was seen as useful, it was adopted; elsewhere it was rejected.

The horse is a good example of an introduced element that was eagerly adopted because, as Roe said, it "enabled the Indians to do *the same things* more easily."[6] Along with the gun, it helped to form the "classical" Plains culture of the early nineteenth century. In many cases, borrowings were not directly from non-Indians; about 1730, for example, a band of Blackfeet with guns encountered a band of Shoshonis with "elk-dogs" (horses). The horse and gun, introduced from opposite directions, had met dramatically.

But not all forms of contact were benign. About 50 years after the above incident, the Blackfeet caught smallpox from their Shoshoni enemies, and more than half the tribe perished.[7] No wonder that when the sky erupted in a spectacular meteor shower in November, 1833, some Indians connected the event with other upsetting changes brought by the White strangers, and asked if the Great Spirit had been unable to keep the universe in order, since the stars were changing places.[8]

And Indians often resisted innovations. As they came to see that European settlement meant the loss of much land and the attempt by missionaries to root out traditional religious practices and replace them with Christian observances, traditional Indians were known to reject the customs and even the foods they associated with the strangers. The Pueblo Indians rose up and expelled their Spanish conquerors in 1680, and were told by their victorious leaders to uproot the Spanish-introduced plants as part of a general return to older ways. U.S. Indian policy derives from the policies of the English, and less directly from those of the other colonizing powers. One must be careful of generalizations about the colonists from different European nations, but it is true that among the Spanish a great number of men came to establish themselves as overlords, to own large ranches and operate mines. They needed Indian labor, so they conquered and reorganized Indian communities for their own purposes. They also felt the need to convert Indians to Christianity and to change any parts of their native cultures which seemed incompatible with that. The French, particularly in the mid-continent, came as traders, who needed the Indians as allies and suppliers of furs, meat, and other necessities. They were often content to coexist with Indians, manage them, and assimilate them as much as possible. But the English typically came as farmers who wanted the land. When Indians proved unadaptable as laborers, they had no place for them in their economic system. Peaceful removal of Indians beyond the frontier seemed the best solution to the English. All the colonial powers recognized the Indians as the true, original owners of the land, and this principle was later accepted by the United States. In spite of that, Europeans failed to understand the Indians' complex relationship to the land, and there were those who denied the Indians' right to what they did not seem to be using as land should be used. Also, all colonial powers insisted that dealings with the Indians, like those with foreign powers, be carried on under the authority of the highest level of government. When the United States came into being, responsibility for Indian affairs was lodged in the federal government, not in the individual states.

The U.S. constitution gave Congress the duty of regulating trade with the Indian tribes. For a time, a "factory system," or series of federal traders supported with public funds, collected more animal pelts than the market demanded. Congress put Indian affairs under the Secretary of the Army, a post first held by John Knox, who advocated keeping the Indians safe on their own land behind recognized borders, supplying them with missionaries, teachers, domestic animals, and ammunition to use in hunting, and prohibiting American citizens from trespassing on Indian lands or driving their livestock there. Congress enacted many of Knox's proposals, but they were for the most part unenforceable. Many American frontiersmen, after the successful war of revolution against a distant king, were not anxious to follow the dictates of a distant federal government that might interfere with their rights, as they saw them, to use land and wildlife as they saw fit, and perhaps to defend those rights by killing anyone, Indian or White, who got in their way.

Gaining land from the Indians by purchase and formal treaty was a major method adopted by the federal government; about 400 treaties were negotiated by the executive branch and ratified by the Senate before the practice was stopped by law in 1871. Treaties generally recognized Indian ownership of land, and conveyed certain parts of that land to the U.S. government in return for goods and services to be paid over a period of time, or in perpetuity. What has never been sufficiently emphasized, however, is the number of treaties in which Indian hunting and fishing rights were retained in the lands to which Indians lost title. Since the European concept of land ownership was difficult for Indians to grasp, and since hunting and fishing were to them the preeminent demonstrations of land use and therefore ownership, it may have seemed to them that they were keeping what was most important. Thus they were encouraged, at least at first, to sign the treaties. To people of European background, title to the land was the essential of ownership, and hunting or fishing were mere usufructs that might later be repudiated by private owners.

For example, the treaties of Medicine Creek (1845) and Point Elliott (1855) in Washington guaranteed the Indians the rights to fish on their "usual and accustomed grounds and stations," and to hunt and gather on "open and unclaimed lands."[9] But the actual effect of these treaties was to remove the Indians from most of their traditional lands and put them on small reservations. At first the Indians had welcomed their new neighbors; as the White patriarch, Ezra Meeker, expressed it: "The Indians, as a class, from the earliest settlements down to the time of making the treaties evinced not only a willingness that the white man should come and enjoy the land with them, but were pleased to have them do so."[10] Joy changed to sorrow when they saw that they would not be left any room in their own land. After the treaties, Chief Leschi, who believed Whites would "sully all the lands and waters of the Indians,"[11] led the Indians of Puget Sound in futile armed resistance. Chief Seattle, a man of peace, had already voiced his unwilling acquiescence:

> But why should we repine? Why should I murmur at the fate of my people? . . . Men come and go like the waves of the sea . . . The very dust under your feet responds more lovingly to our footsteps than to yours, because it is the ashes of our ancestors, and our bare feet are conscious of the sympathetic touch, for the soil is rich with the life of our kindred.[12]

What Indians were losing was not only their economic base, but their spiritual connection with the ancestral land.

Thus, Indians were dispossessed of their lands through treaties, sale and other arrangements. Maps of the acquisition of land by the United States almost always show large blocks of territory purchased or won by treaty from Britain, France, or Mexico. But in actuality the land was taken by conquest, purchase, and treaty from its Indian owners. It would be interesting to see a map detailing that process, with the names of the original owners and the dates on which the land passed out of their control.[13]

Treaties often took only part of a tribe's land and left the Indians a remnant on which to live, but in the 1820s and 1830s, a policy of total removal to other lands west of the Mississippi River was enforced. Some of the tribes that had taken the furthest steps to adopt White civilization, such as the Cherokees, were the first to be driven west on foot and in carts to what is now Oklahoma. In addition to the terrible loss of life — about one quarter died along the way — the ecological dislocation involved in moving from the Southeastern forests to the Great Plains was a major challenge. Similar blows fell on tribes located in the South, the Ohio Valley and the Great Lakes region. Uprooted from one ecosystem and dropped into another, they were given little

CHAPTER IX

time to make the necessary adjustments before the frontier reached their new lands. For a few tribes, the process was repeated more than once. Indian removal was justified by saying that the Indians were not fully using the land, and should make way for another people who knew how to till the soil and support a larger population. But the tribes first removed were the "civilized tribes" of the South like the Cherokees, whose farming methods were in no way inferior to those of their White neighbors, and who had a written language, newspapers, a constitution, courts, and schools as well. The real reason for their removal was that more powerful people coveted their lands, where gold had been discovered not long before. This statement is supported by abundant evidence, including a Georgia law passed in 1830, eight years before the removal, that among other things prohibited the Cherokees from mining gold on their own land.

In the course of the nineteenth century some, but by no means all, American citizens desired not merely the removal of the American Indians to distant territories, but their complete physical destruction. These people used as their motto, "The only good Indian is a dead Indian." Their program was not ultimately successful, but it made shocking inroads as greedy, lawless whites took matters on the frontier into their own hands, and as the armed forces of the United States were used to pacify, defeat, and in many cases to slaughter Indians. Sometimes the Army used deliberate ecological warfare against the Indians by destroying the animals and plants on which they depended. When Kit Carson defeated the Navajos in 1863 before taking them into captivity at Fort Sumner, New Mexico, he used "scorched-earth" tactics, slaughtering their herds, burning their fields and stores, and cutting down their fruit trees, as well as shooting all the wildlife in sight.

The most widespread and catastrophic use of environmental tactics against Indians was the virtual extinction of buffalo and the serious reduction of other game animals on the Great Plains. The slaughter of the buffalo was done partly for profit, and partly as a deliberate act of ecological warfare against the Indians. Officals in Washington were convinced that the Plains Indians could not be induced to live a settled, civilized life while the buffalo herds were accessible. The feelings of some military campaigners against the tribes were echoed in the legends stamped on proprietary coins circulating in the west: "One dead buffalo: One dead Indian: One dollar." While the herds were falling in thousands before White men's guns, it is not surprising that some Indians abandoned the older practices of conservation and killed as many buffalo as they wanted, disregarding their elders' pleas and admonitions, since they could see that if they did not do so, the White men would shoot them anyway. Others were willing to hunt buffalo, taking only the tongues, for the Whites after they had been corrupted by trade goods and liquor. It was difficult for Indians to envision the eventual extinction of the buffalo, but some did try to protect them. During twelve years or so before 1876, there seems to have been a concerted effort to keep the buffalo herded back out of the area where Whites were hunting them most heavily.[14] One of the remnant herds, about thirty animals, was saved by Walking Coyote in the center of the Flathead Indian Reservation, in an area that later was made the National Bison Range.[15] The southern plains were empty of buffalo by 1874, and the northern herd was almost completely annihilated in nine more years. A few scattered groups remained in isolated places, mostly in the mountains, but the days of abundance were over. The severity of the shock for the Plains Indians was expressed by Thomas J. Morgan, U.S. Commissioner of Indian Affairs under Benjamin Harrison:

> The buffalo had gone and the Sioux had left to them alkali land and government rations. It is hard to overestimate the magnitude of the calamity, as they viewed it, which happened to these people by the sudden disappearance of the buffalo and the large diminution in the numbers of deer and other wild animals. Suddenly, almost without warning, they were expected at once and without previous training to settle down to the pursuits of agriculture in a land largely unfitted for such use. The freedom of the chase was to be exchanged for the idleness of the camp. The boundless range was to be abandoned for the circumscribed reservation, and abundance of plenty to be supplanted by limited and decreasing government subsistence and supplies. Under these circumstances it is not in human nature not to be discontented and restless, even turbulent and violent.[16]

Adding insult to injury, the non-Indians blamed the Indians for killing off the buffalo by drives and other wasteful hunting methods. But the Indians themselves were not responsible for this misfortune. As we have explained, their actions in respect to buffalo were consistent with the preservation of the species. The "wasteful buffalo Indian" never existed in reality, but was one of the stock accusations made by anti-Indian writers during the wars of the second half of the nineteenth century. But a "careful and critical examination of their arguments yields

CHAPTER IX

some curious results in certain cases, both in respect of polluted channels of information, and of sweeping assertion unsupported by adequate evidence; this last being evidently deemed superfluous when the accused parties were only Indians!"[17] The "wasteful Indian," the "rootless nomad," and the "only good Indian" ("One who was dead") are all products of the same deprecating attitude that was so pervasive at that time that even such an enlightened and sympathetic writer as James Mooney could let slip an expression like this one: "Without the horse the Indian was a half-starved skulker in the timber, creeping up on foot towards the unwary deer."[18]

The fact that the buffalo Indians were "nomads" in the sense of having movable dwellings and following the herds does not mean they had no attachment to the land. They had recognized territories within which they traveled, and they loved the homeland. The real "rootless nomads" on the Great Plains were White commercial hunters, trappers, traders and tourists who came, killed the animals, and left. They were the ones who had no attachment to the land, and it was they who depleted the plains. It was noted in 1858 "that the disappearance of the large quantities of game has only taken place within the last twenty years."[19] It was to accelerate almost beyond measure in the twenty years following. Initially, Indian attitudes toward nature were strong enough to prevent them from assisting the White raid on animal resources in some ways. The Blackfeet would not kill the plentiful beaver to trade with the Hudson's Bay and North West Companies because the beaver was a sacred animal.[20] They did trade buffalo meat, of which they had a good supply. But attitudes eroded under constant pressure.

Widespread alterations in the ecosystems that supported Indians had similar effects even when they were not the results of premeditated policy. Trappers sought out and removed fur-bearing mammals from millions of acres; commercial hunters severely depleted many kinds of wild birds, and brought some, like the sky-darkening hordes of passenger pigeons, to extinction. This process of the destruction of wildlife, which occurred everywhere, can well be illustrated by the historical experience of the Northwest Coast Indians.

The Northwest was geographically isolated from Europe and the United States, so that the first recorded visits of Spanish, British, American, and Russian ships date from the late eighteenth century. Their first concern was to develop the fur trade, and they were all too successful with the first target species, the sea otter. The sea otter trade was profitable, because pelts could be bought from the Indians for a tiny fraction of the price they would bring on the market in China, Europe and eastern America. Enterpreneurs brought in their own hunters as well, and by 1825 the sea otter had been so thoroughly decimated that it was believed to be extinct. Then the traders turned to whale oil and the pelts of bear, marten, and land otters. Many times, especially in the last-named case, they had to overcome an Indian inhibition against hunting certain animals, but they did so by offering a whole series of new products that were irresistible within the context of the Indian's acquisitive and gift-giving culture: carving tools, kettles and spoons, bars of iron and sheets of copper, guns and traps, cloth, blankets, potatoes, tobacco, and alcohol. The last was a mixed blessing, of course. At the same time, venereal and other diseases were being transmitted. A terrible outbreak of smallpox wiped out whole villages in the 1830s.

There was some cultural resistance to the introduced products, expressed in terms of religious prohibitions. It was believed that the eulachon fish refused to be boiled in metal kettles, or anything other than the traditional wooden containers. Salmon, it was said, would not return if they were cut with the White men's knives. But the items gained through trade soon made their appearance at the potlatches; eventually, trade blankets were the commodity most commonly given at feasts.

Sometimes the European-American onslaught met with determined opposition. The Tlingit Indians burned the Russian outpost at Sitka to the ground in 1801, two years after it had been first established. Several British and American ships were captured by the Nootka and others. But the Indians were terribly vulnerable. When a White force too strong for them arrived, they could flee safely into the forests, but their possessions could be destroyed, and without houses, canoes, smokehouses, almost all their fishing and hunting equipment, and their stores for the coming year, they were virtually helpless.[21]

Whalers moved into the waters off the Northwest Coast around 1800. Earlier visitors had been awed by the numbers of whales to be seen there; a reliable account from 1785 reports that twenty or thirty were always in view.[22] American whaling had its heyday in the decades just before the Civil War, and the huge mammals were rare in Northwest waters by 1874. Indian whaling ceased in the early 1900s. Meanwhile, with the land mammals depleted and the sea otters and whales practically gone, White commercial interests concentrated on seals.

CHAPTER IX

Wholesale slaughter of the fur seals was underway in the 1870s and 1880s; in those two decades, two and a quarter million seals are known to have been taken.[23] In twenty more years, the seals were almost gone, and in 1911, in a rare moment of international agreement on conservation, a fur-sealing convention was signed with limits on the numbers that could be killed in the few remaining herds.

The Indians had seen the species they had hunted decline or disappear on land and sea. Older Indians surmised that the animal beings fled from the sound of guns, and that young men were no longer making the effort to purify themselves before the hunt.[24] But one source of abundance, fishing, especially the salmon, still remained, and the Northwest Coast Indians had managed to hang on to that remnant, at least, of their traditional ecological relationship, even if they were beginning to lose the ceremonial life that gave it meaning and assured its preservation. And it seemed that the salmon might be the next species to go. Following the decline in sealing, major commercial development of the Northwest Coast fisheries became possible through improved transportation and methods of canning. While commercial fishermen moved into Indian fishing grounds in a big way, they did provide wage-earning jobs on the ships and in the new rendering plants and canneries, and Northwest Coast Indians were prepared through their own culture to take these jobs. More traditional Indian fishing methods persisted, but the places they could be used were fewer, and the numbers of fish, too, were declining. In a few pages, the story of Northwest Coast fishing rights will be resumed in a recent context. But let us return to the more general narrative.

Indian agriculture and pastoralism also felt the effects of the American conquest. In area after area, squatters settled the best land. Clearing for agriculture and wasteful logging and mining practices removed whole forests from the face of the continent and accelerated erosion of the soil. Even in terms of their own future needs, the pioneers wasted resources prodigally; in terms of the native inhabitants, what they did was to destroy cultures and peoples by shattering their ecological bases.[25] At the same time, tribes were still losing hundreds of people to epidemic diseases. Sometimes these were introduced deliberately, as when blankets that had been used by smallpox patients were given by soldiers to the Pawnees, but more often they spread by themselves, as Indians with no inherited or developed resistance came into contact with populations harboring the agents of infection.

The Strangers' Ways

The administration of U.S. Indian policy was removed from the War Department and given to a Bureau of Indian Affairs in the newly created Department of the Interior in 1849, a governmental body whose responsibilities lay chiefly in the area of public lands and natural resources. The federal policy from 1860 to 1890 in a large section of the country was to confine the remaining Indian population on reservations, either by removal to the Indian Territory in Oklahoma, or by setting aside a portion of the traditional tribal lands. Reservations were under the supervision of the Bureau of Indian Affairs. "Friendly" Indians were promised protection and support if they settled down on the reservations; those who stayed outside were automatically defined as "hostile," and either rounded up or hunted down by the Army. Reservations were, almost by definition, fragments of a former ecological unity. They did not contain enough wild game to support the more concentrated populations located on them, and Indians were almost never permitted to leave the reservations to hunt. The hunting grounds were thus depleted, or closed to Indians, or both. Indians lived on government rations that were diminished, or delayed, and sometimes failed to arrive at all. Friends and foes of the Indians alike said that the hunting, fishing and gathering tribes could no longer survive as they were. They would have to to change their way of life, or die. A U.S. senator described the situation as he saw it in 1881:

> Our villages now dot their prairies; our cities are built upon their plains; our miners climb their mountains and seek the recesses of their gulches; our telegraphs and railroads and post offices penetrate their country in every direction; their forests are cleared and their prairies are plowed and their wildernesses are opened up. The Indians cannot fish and hunt. They must either change their mode of life or they must die. That is the alternative presented. There is none other. We may regret it, we may wish it were otherwise, our sentiments of humanity may be shocked by the alternative, but we cannot shut our eyes to the fact that that is the alternative, and that these Indians must either change their modes of life or they will be exterminated.[26]

The preferred alternative to extinction, then, was assimilation. Programs would be developed to break down the traditional patterns of Indian life and make them more like White American citizens. In the process, their sacred traditions were to be forgotten, their views of nature replaced by the teachings of the dominating society's education

CHAPTER IX

and religion, and their ways of relating to the natural environment replaced by those of their conquerors. Indians would have to learn to work steadily and hard. Hunting tribes would have to try to learn farming; men in tribes where the women farmed with digging sticks would have to learn to walk behind the plow; and even some Pueblo Indians, who retained their original farmlands, would still have to send their children to school to learn the White ways. Assimilation was conceived not only as a means of survival for the Indian, but also as an end in itself, something that would be as good for society as the Americanization of the multitudes of European immigrants who were then arriving to help fill up an "empty" continent. If successful, it would mean the end of traditional Indian environmental attitudes and practices.

Education, for the Indians, was traditionally left in the hands of the family and community. What was provided for them by the dominating society was institutionalization. In the early years of the nineteenth century, the U.S. government encouraged and supported missionary schools for Indians, where Christianization was as important a goal as basic instruction. Missionaries taught the Indians that their old ways of life were false and outmoded. William Duncan, for example, came to the Tsimshian at Fort Simpson in 1857, and moved "his Indians" twice to two settlements, both named Metlakatla, to take them out of their old culture and away from the corrupting influences of White neighbors at the same time. Unlike most other missionaries, Duncan took the trouble to learn the local language and to provide for the economic survival of his community. But there were no totem poles or other carvings in Metlakatla, and elsewhere they were burned, or abandoned, or collected. Indians soon learned that any time they expressed a view of nature as possessing spiritual power, they would be corrected, always with disapproval and often with scorn, ridicule, or punishments. By the 1880s, compulsory government schools were replacing missions as the usual places for Indian education. Around the turn of the century, the Bureau of Indian affairs had assumed responsibility in this sphere, and had embarked on a program of education for Indians that was designed for their acculturation. Federal educators favored boarding schools located far away from the reservation, so that children could be removed from their families, and therefore from the influences of older ideas and ways. They were taken hundreds of miles from the rich interrelationships of their native environments and put into rigidly structured dormitories and classrooms. They were strictly forbidden to speak their native languages which contained so many natural associations and an implicit view of the world. Whatever was Indian was, in almost all these schools, something to be ashamed of, to be laughed at, to be punished for. If any Indian ecological understandings survived such schools, as some did, the fact of that survival is miraculous.

It is important to remember that acculturation was not just a process of education, but a total assault on the Indian way of life, using resources of government and aided by many of the economic and social activities of non-Indians. The Plains tribes confined to reservations and forced to become farmers, are a case in point. Even those who had been used to raising maize found the problems intolerable. Women, traditionally their agriculturists, were now supposed to say indoors while men went out to do what had been women's work. Instead of the digging stick, they were told to lacerate Mother Earth's breast with the plow. Instead of sharing with the extended family and other members of the group, they were urged to work for themselves and "save." Their children were taken away to be given an alien education, and they were to be taught that their old religion of honor for nature was ridiculous and mistaken. Treaties guaranteed their rights to hunt in their "accustomed territories," but they were to have no opportunity to do so, and anyway there was nothing to hunt. Meanwhile, not only the buffalo had gone. In the Black Hills, the sacred land of the Sioux and Cheyenne, hastily built wooden towns appeared and mine tailings spilled down the hillsides. The Teton Sioux men, Shooter and Higheagle, observed, "The water of the Missouri River is not pure, as it used to be, and many of the creeks are no longer good for us to drink."[27]

In the Northwest, active government intervention in Indian culture also intensified. Shamans, raiding, and potlatch customs met with disapproval. The potlatch was prohibited from the 1880s well into the twentieth century, with little understanding of its place in Indian societies. Alcoholism became a serious problem as Indians saw their way of life being smashed to bits. Prohibition of trading alcohol with Indians produced a new problem: the smuggling of *hoochinoo*, a drink fermented from a mixture of such ingredients as flour, molasses, dried applies, salmonberries, blackberries, and water. The word was adopted into English as "hooch."[28]

Southwestern Indians, who had generally managed to retain their traditional attitudes toward nature through the Spanish period, also resisted the changes urged by the Anglo-Americans, and preserved

CHAPTER IX

many of their attitudes and customs. Mechanization of agriculture came comparatively late to the Pueblo Indians, and met with some resistance as a departure from custom and an insult to Mother Earth. A man of Jemez Pueblo was taken from his field by the town elders and whipped for plowing with a tractor not long before 1939.[29]

During the spring, when plants are starting to grow, the Pueblo Indians believe the earth is tender and treat her with special care. Often they remove the heels from their shoes and the shoes from their horses' hooves during that season. Barre Toelken reports that he once asked a Hopi "'Do you mean to say, then, that if I kick the ground with my foot, it will botch everything up, so nothing will grow?' He said, 'Well, I don't know whether that would happen or not, but it would just really show what kind of a person you are.'"[30]

Christianization was either rejected, as among the Hopis, or the new religion became a facade beneath which the older beliefs and ceremonies functioned in Christian guise, or were continued in secret.

But everywhere in America, the process of acculturation was accompanied by an attempt to Christianize Indians and to stamp out tribal religions. The previous chapters have explained how deeply interwoven Indian religion and ecology are, an attack on one was inevitably an attack on the other. The Sun Dance and other public ceremonies were forbidden. Sacred objects had to be hidden from the authorities. In 1884 a federal regulation made the practice of any native American Indian religion illegal. But this policy was met by determined resistance by many Indians. In the later half of the nineteenth century, some native religions arose that spread beyond tribal boundaries, even in the face of strong governmental opposition.

About 1860, the Indian prophet Smohalla had a vision in which a mountain spoke to him, and he began to preach a revival of old ways on the Columbia River Plateau, beyond the northwestern margin of the Great Plains. He told his followers to give up all White ways, saying that hunting and fishing were "natural work" that did Indians no harm, but that one who does the White man's work "cannot dream."[31] In explaining his opposition to the Indian Homestead law, Smohalla made clear his desire to restore a proper relationship between the Indians and the natural environment:

> God . . . commanded that the lands and fisheries should be common to all who lived upon them; that they were never to be marked off or divided, but that the people should en-

joy the fruits that God planted in the land, and the animals that lived upon it, and the fishes in the water. God said he was the father and the earth was the mother of mankind; that nature was the law; that the animals, and fish, and plants obeyed nature, and that man only was sinful. This is the old law. . . .

It is a bad word that comes from Washington. It is not a good law that would take my people away from me to make them sin against the laws of God.

You ask me to plow the ground! Shall I take a knife and tear my mother's bosom? Then when I die she will not take me to her bosom to rest.

You ask me to dig for stone! Shall I dig under her skin for her bones? Then when I die I cannot enter her body to be born again.

You ask me to cut grass and make hay and sell it, and be rich like white men! But how dare I cut off my mother's hair?

It is a bad law, and my people cannot obey it. I want my people to stay with me here. All the dead men will come to life again. Their spirits will come to their bodies again. We must wait here in the homes of our fathers and be ready to meet them in the bosom of our mother.[32]

Smohalla's movement spread to many tribes in the northwestern states, but as an Indian ecological protest movement, it was eclipsed by the far more widespread Ghost Dance between 1870 and 1890.

The Ghost Dance, which promised the return of the buffalo and other game, the restoration of traditional ways of life, and the disappearance of White people and all their works, began among the Paiutes of Nevada. Many tribal religions contained the idea of earth renewal, associated originally with the revival of all life in the springtime, a concept which the Ghost Dance extended to include the resurrection of the Indian dead and all the wild animals that had disappeared. This would happen if the Indians believed, revived the old ways, danced a round dance and sang powerful songs. A prophet, Wovoka, began to breathe new life into the movement in 1889, and it spread quickly across the plains, where men like Sitting Bull were unwilling to accept the dual destruction of their natural environment and their traditional way of life. "God Almighty has always raised me buffalo to live on,"[33] he said, and it seemed inconceivable that God should then have allowed all the buffalo to be destroyed. Many Plains Indians believed that a new earth, a paradise of long grass, abundant plants, and great herds of wild game, would come sliding over the old earth, bearing with it the dead returned to life. Indians would be lifted up onto the new earth to live forever, but Whites would disappear with the old. The restoration of harmony with nature was a major theme of the Ghost Dance. Wovoka, it was said, could make animals talk, and he taught a way of treating the remains of a buffalo so that it would come back to life after it had been butchered. Short Bull and Kicking Bear, respected Sioux men who had been sent to visit Wovoka, returned with the description of a vision they had been granted:

> A big eagle carried them over the clouds to a happy place. . . . Everything there was beautiful. Nothing could be seen that was of the White man's making. . . . Around them, as far as the eye could see, were herds of elk and buffalo. The Great Spirit told them that the earth was old and worn out, ruined by the White man's greed. If the tribes would follow Wovoka's instructions, the beautiful land would become a reality and the dead would return.[34]

Many of the Ghost Dance songs promised the renewal of natural abundance:

> Over the glad new earth they are coming
> Our dead come driving the elk and the deer.
> See them hurrying the herds of buffalo![35]

A Sioux song said that the message of hope was brought by the sacred birds:

> The whole world is coming,
> A nation is coming, a nation is coming,
> The eagle has brought the message to the tribe.
> The father says so, the father says so.
> Over the whole earth they are coming.
> The buffalo are coming, the buffalo are coming,
> The crow has brought the message to the tribe,
> The father says so, the father says so.[36]

And a Kiowa Ghost Dance text, sung with repetitions to a very good tune, expressed an aspect of the Indian environmental attitude:

> Because I am poor,
> I pray for every living creature.[37]

Several tribes placed the sacred cedar tree in the center of the Ghost Dance circle, thus echoing older traditions.

The Ghost Dance met with the determined hostility of U.S. Indian Agents and the U.S. Army almost everywhere it appeared. The first incident at Wounded Knee, in December, 1890, marked the climax and began the end of the Ghost Dance's popularity. There scores of Sioux believers, including women and children, were killed by soldiers. The promised new world did not appear, and songs and other remnants of the Ghost Dance were blended into other tribal renewal ceremonies like the Sun Dance. Indian longing for the reestablishment of integrity in their relationship to the earth and its creatures cannot be said to have disappeared, but even the eager expectation of that had to be replaced by disappointment and patience. The feeling remained: "A man ought to desire that which is genuine instead of that which is artificial."[38]

The most successful of the new Pan-Indian religions in reaching every part of Indian America and surviving, as it did, to remain a potent religious force in the present day, is the Native American Church. Beginning in pre-Columbian northern Mexico, it spread first north of the Rio Grande among the tribes of the southern Great Plains as the Ghost Dance was ending. It incorporated some Christian teachings, and made the experience of living in two vastly different cultures easier and more comprehensible for those who "followed the Peyote road." It is significant that the Native American Church retained a sense of the power and importance of the natural world, and gave a plant, the peyote cactus, a central place in ceremony, sacrament, prayer, and story. The use of peyote became the issue in persecution of the Native American Church. Although a federal law was never enacted against it, state after state prohibited its use as a "dangerous drug," in spite of the fact that it is not addictive in the natural form in which it is consumed, and its use is regulated by the social setting provided by the ceremony itself. Such laws have more recently been held to be in conflict with the First Ammendment right of freedom of religion in the U.S. Constitution. By the middle of the twentieth century, more than half of all American Indians felt a sense of identification with this Pan-Indian religion.

No sooner had the system of Indian reservations taken shape than voices were raised for their fragmentation. If Indians were to be assimilated, it was said, they ought not to be kept on large pieces of land held in common by the tribe. They should be given pieces of land

CHAPTER IX

in individual private ownership, and taught the methods of agriculture and the value of labor. They would need supervision for a while, but eventually — 25 years was the hope — would be self-sufficient farmers like other Americans, and could be handed deeds to their land and made citizens. This was the avowed intent of the Indian General Allotment Act of 1887. It provided for the assignment of tracts of reservation land, usually 160 acres, to individual Indians. When all the recognized individuals in a tribe had been taken care of, the remaining land was declared "surplus," bought from the tribe, and opened to non-Indian settlement like public land under the Homestead Act. The "surplus" land was usually the largest part of the reservation. Indian allotments did not necessarily adjoin each other; sometimes they were deliberately arranged in a "checkerboard" fashion so that Indians could live side by side with the new White neighbors and learn from them. The situation was ready made for non-Indian land grabbers, who moved in to take the more desirable sections of "surplus" land, and to gain control of the Indian allotments through lease, purchase, or theft. Congress acted to assert the status of the federal government as trustee of Indian lands, and to prohibit the sale of allotments within the 25-year period, but the loss of Indian lands continued at a rapid pace for almost 50 years. Two-thirds of all lands belonging to Indians in 1887 were in non-Indian hands by 1934. The result was to fragment such ecological integrity as had remained on the reservations, and to encourage a raid on Indian resources.

Fortunately, a few reservations, mostly in the Southwest, escaped the allotment process, probably because they were too isolated and arid to excite much non-Indian greed and its attendant political pressure at that time.

President Theodore Roosevelt, aware that deals were being made by unscrupulous companies to cut Indian timber, destroying whole forests with no provision for sustained yield, tried to place most Indian forest lands within the National Forest system. At the time, this would have provided protection for the forests but would have reduced further the lands included in Indian reservations. In most cases, Roosevelt's attempts failed, and most Indian forest lands then existing on reservations were not included in National Forests. Of course, all National Forests and National Parks, as well as all private lands, are located on what was Indian land at one time.

By the 1920s, it was evident to many observers that the land allotment and assimilation policies of the U.S. government had failed.

Indians had lost much of their land and many of their ways of life without becoming economically self-sufficient or able to function within the larger society, except for a distinguished few. In 1924, all Indians were made citizens. Four years later, a private research group under government contract issued an influential report highly critical of U.S. Indian policy and its administration by the Bureau of Indian Affairs, and urging reform.

Far-reaching reforms began in earnest under John Collier, appointed U.S. Commissioner of Indian Affairs by Franklin Roosevelt in 1933. Collier was a social scientist who respected Indian cultures and was sensitive to Indian ecological perspectives as few non-Indians have ever been. Instead of attempting to assimilate Indians to non-Indian culture, he believed that Indians had cultural contributions to make to the totality of American life. One of the most important of these contributions, he said, was "that passion and reverence for human personality and for the web of life and the earth which the American Indians have tended as a central, sacred fire."[39] Collier voiced his own surprise at the extent to which Indian culture and ecological perspectives survived in his own day, particularly in the Southwest.

Collier worked actively to reform the Bureau of Indian Affairs, and many of the changes he recommended were included in the Indian Reorganization Act, passed by Congress in 1934. The allotment of Indian lands was stopped, and those portions of the lands in Indian Reservations that had been declared "surplus," but had never passed into private hands, were returned to tribal ownership. Federal trust protection was reasserted over Indian lands. Some funds were appropriated to help tribes buy back land needed to consolidate their reservations. For a period of time, then, the Indian land base stopped diminishing, and actually began to grow. Unfortunately the alienation of Indian lands did not stop, however, and in the period 1934-1974, 1,811,010 acres were lost, including 488,226 acres for dams, reservoirs, and other water projects.[40]

In the 1930s, the government began several programs on Indian Reservations to deal with problems of soil, water, and forest conservation. The Bureau of Indian Affairs located in the Department of the Interior with its orientation toward natural resources, was responsible for the development or safeguarding of Indian properties such as land, oil and gas, water rights, forests, and fisheries. The Bureau had, and still retains, supervision of annual timber sales worth millions of dollars. Many Indian tribes developed well-trained firefighting crews that

CHAPTER IX

became justly famous as they were called upon to fight fires in National Forests far from their own reservations. An Indian branch of the Civilian Conservation Corps was organized in 1933. Several irrigation projects were initiated in this period.

The Indian Reorganization Act allowed Indian tribes to organize their own tribal governments, not along traditional lines, but after the model of U.S. municipalities, with constitutions and elected councils. Tribes were also permitted to incorporate to manage their funds and begin to gain a degree of control of their own resources, although federal supervision was ever-present. Some tribes began their own businesses, such as cattle cooperatives and fish canneries.[41] A few tribes received large incomes as their mineral resources, particularly fossil fuels and metallic ores, were exploited.

Many of the reforms of the Collier period were salutary and overdue. The extension of freedom of religion to Indians is an excellent example, and the encouragement and recognition given to traditional Indian arts is another. The recovery of large tracts of Indian land for the tribes was a major accomplishment. But ecologically, the results were mixed. Though Collier appreciated the fact that Indians were our first conservationists, Indians were not given control in the planning and execution of projects on their own reservations. Conservation of Indian resources was not developed from the Indians' own environmental perspectives, but was the economically motivated conservation of early twentieth-century America imposed upon the Indians. No matter how necessary and valuable were some of the steps initiated for the Indians by the federal government, they were often done in ways that further disrupted Indian patterns. Perhaps the best-known example of this was the administration of the stock reduction program on the huge reservation of the Navajo Indians. Following the Treaty of 1868, they had returned to their homeland to live in peace, doing everything possible to increase the herds of domestic animals that were their pride and sustenance. Soon the numbers of animals exceeded the carrying capacity of the land; vast tracts were stripped of useful plant cover, and subjected to severe erosion. A critical point was reached in the last quarter of the nineteenth century and the first quarter of the twentieth, with deeply entrenching arroyos depriving fields of water access.[42] In the 1930s, the paternalistic B.I.A. under Collier enforced a stock reduction program that was certainly needed from an ecological standpoint but handled badly from the standpoint of human relations, as Collier himself later admitted.[43] To some Navajos, watching the forced sale or slaughter of their beloved animals, it seemed as if some of Kit Carson's methods had been revived.

After the Second World War, a different spirit animated U.S. Indian policy. It was aimed at terminating the federal responsibility for Indians. The government, through resolutions in Congress and several new laws, declared itself ready to end special status for Indians, tribe by tribe, leaving them to sink or swim in the American mainstream. Only a few tribes were actually terminated, but the ecological results were disastrous for those tribes, as federal protection for their lands, forests, and other resources were withdrawn.

The Klamath Reservation in Oregon consisted of thousands of acres of mainly forested land. When they were terminated, about 80% of the land was sold to the U.S. Forest Service and private companies, and the money was distributed among the 80% of the tribe that had decided to accept termination. The other 20% of the tribe, called the Remaining Group, kept the rest of the land, and only they were permitted to use it. A court decision established that the terminated Klamaths were no longer Indian under federal definition, and could not hunt or fish on the land of the Remaining Group.[44] More recently, the Remaining Group also sold its land, so that the relationship of the Klamaths to their ancestral forest has been ended.

When the Menominees of Wisconsin were terminated in 1961, they kept their land, but as a private corporation rather than an Indian tribe. Their reservation became the poorest county in Wisconsin, and because their forest products industry could not support them economically, the corporation began to sell vacation homesites around one of the prettier lakes in Menominee County, Spirit Lake. Again, the Indian land base and Indian control over their own resources seemed threatened. A strong lobbying effort in Congress succeeded in restoring the Menominees to federal trust status, which prevents the further dissipation of land, reestablishes federal services and makes the Menominees officially Indians again.

The Indian Claims Commission was created in 1946 to handle cases in which Indian tribes had lost their lands without fair payment, and to decide the amounts of awards which should be made to rectify those inequities. In short, Indian lands were not to be returned, but descendants of defrauded ancestors were to be paid for the lost land. Many awards, some of them amounting to millions of dollars, have been made by the Commission.

CHAPTER IX

During the 1950s, a federal program was initiated to help Indians move from reservations into cities. This program, usually called "relocation," was administered by the Bureau of Indian Affairs.[45] It provided selected Indians with transportation, living expenses, job training, and help in finding employment, for a limited period of time. Though relocation was not part of termination, many Indians saw it as another way of ending federal responsibilities. At the same time, thousands of Indians were moving to the cities on their own. The result was that the Indian population, which could be described as predominantly rural before this period, became half composed of urban residents, although most kept strong family ties with the reservation. Most reservations did not lose population because of the high Indian birth rate and the fact that health services, while still substandard, were improving. An increasingly urban Indian population came into contact with environmental problems familiar to most Americans.

Indian reservations were often invaded by dams and reservoirs that flooded their best lands. Although payments were usually made for inundated land, the disruption of tribal life and their relationship to the natural environment is hard to measure. More serious than any other single cause in the depletion of the salmon runs in the rivers of the Northwest has been the construction of power-generating dams. Grand Coulee Dam alone made hundreds of miles of spawning grounds useless to the salmon. Even dams provided with fish ladders and other means of getting the salmon around the barriers take their toll. The traditional Indian fishing grounds at Celilo Falls on the Columbia River were drowned in a reservoir. In the Northeast, the Senecas fought the construction of Kinzua Dam on the Allegheny River to the bitter end, but lost. The Ft. Berthold reservation in North Dakota was divided into several disconnected pieces by a reservoir. More recently, a political battle raged around the authorization of Tellico Dam, which would flood historic Cherokee land and threaten the extinction of the Snail Darter, a species of fish. Opposing the dam, a Cherokee named Jimmie Durham spoke before a congressional committee as a representative of the International Treaty Organization. Appalled by the attitudes of many politicians who would sacrifice land and fish to generate electricity, he said,

> "It is this incredible arrogance towards other life that has caused such destruction in this country." "Who," he asked, "has the right to play God and judge the life or death of an

entire species of fellow beings which was put here by the same power that put us here? Who has the right to destroy a species of life, and what can assuming that right mean?

Let me be emotional: To me, that fish is not just an abstract "endangered species" although it is that. It is a Cherokee fish and I am its brother.[46]

In this case the Indians lost again. The dam was built and energy-generating technology triumphed over the concept that land and living things are sacred.

Indian water rights have not always been well protected by those who had the authority to do so. Water reclamation projects were seldom developed with Indian needs and entitlements in mind; tribes like the Pimas were deprived of their accustomed sources, and other projects sometimes failed to deliver promised water. Several recent controversies have centered around this issue. For many years, the Paiutes saw the water of Pyramid Lake, Nevada, sink lower and lower, due to irrigation diversions of White agricultural businesses upstream. The Indians' fish and waterfowl were endangered. Their neighbors, the Washoes, had their major stream polluted by sewage from the rapidly growing towns around Lake Tahoe. Waterfront usage has become an issue along the Washington coast and elsewhere, and Indian rights in the tidelands and offshore waters seem sure to be asserted increasingly.

Another area of conflict and litigation is that of Indian rights to fish and wildlife. Here again, the contrasting attitudes of Indians and others caused problems in earlier days. Indians noticed the strangers' marked lack of respect for wild animals, and sometimes took advantage of it. A White man might be asked to kill a troublesome, cornfield-raiding bear, thus saving the Indians from the wrath of the bear spirit. As White commercial hunters began to decimate the wild herds, Indians believed they knew the reason: Whites did not do the proper rituals, or apologize to the animals' spirits, so once killed by them, the animals did not come back to life again. New weapons changed Indian hunting patterns, and other pressures modified attitudes as well. Today mountain sheep, antelope, and deer are rare in many reservation areas. Some Indians now speak of hunting as "fun," at least to Whites, or serve as guides for White hunters, but the traditional attitudes are not dead everywhere. If an Indian kills a wild animal, he can usually give a good reason for it. But the older patterns are fragmentary because the

CHAPTER IX

environment itself has changed, and more serious changes still threaten.

A well publicized issue in recent years was the question of Indian fishing rights in the State of Washington. Even though these rights were guaranteed in the treaties, as early as 1912 the state prevented Indians from fishing in some streams because, as non-citizens, they could not get fishing licenses. An Indian agent commented on the situation:

> The natives' natural larders have been chiefly the shellfish and fishery locations adjacent to the mouths of the great rivers of this vicinity. These resources have been sufficient to subsist and maintain our Indian people hitherto. These resources have naturally lessened with the advent of the white man; more recently, the use of large capital, mechanized assistance, numerous great traps, canneries, etc., and other activities allied to the fishery industry, have greatly lessened and depleted the Indians' natural sources of food supply. In addition thereto the stringent and harsh application to Indians of the State game and fish laws have made it still and increasingly precarious for him to procure his natural foods in his natural way. Much of this had been done under color of law. An empty larder, however, is an empty larder.[47]

Some temporary relief occurred when Indians were declared citizens in 1924. But the basic problem remained and was to re-emerge in another form.

Declining numbers of fish, due to the use of new technology in locating, catching, and freezing by commercial fishermen, and the accelerated efforts of foreign fishing fleets, brought about another crisis for the Northwest Coast Indians in the years after the Second World War, affecting their one major surviving traditional economic activity. Government regulations, including licensing, closed seasons, and closed waters, were more widely enforced. In spite of the treaty guarantees of Indian fishing rights, the State of Washington enforced its own conservation regulations against the Indians, and provoked the first widely publicized demonstrations of Indians in the latter half of the twentieth century, the "Fish-Ins."

Indian organization in defense of their own rights was not a new idea along the Northwest Coast. Beginning in 1912, the Alaska Indians had formed the Alaska Native Brotherhood and Sisterhood, followed by similar organization in British Columbia in 1931. These early bodies petitioned for the recognition of Indian land and fishing rights, but also accepted assimilation and the abolition of "aboriginal customs" as a stated goal. But the new movements of the 1960s, the Survival of American Indians Association and its allies like the National Indian Youth Council, combined defense of Indian rights with a deeply felt desire to preserve and even revive traditional Indian culture, including fishing, which is so important a part of the spiritual as well as economic way of life for the Northwest Coast Indians. Indians went to fish at their "usual and accustomed grounds and stations," in violation of State regulations but in accord with the treaties, and many of them and their allies were arrested, in 1964 and the years immediately following. In 1966 they received the aid of the Department of Justice, and in a long series of court cases, Indian fishing rights under the treaties were reaffirmed. The State is now required to protect those rights and it is up to the State to "establish that it is necessary for conservation to impose the specifically described restrictions on the exercise of treaty rights."[48]

In the last two decades, tribes have gained more control over their own wild animals and plants, as tribal councils have begun to set seasons and issue hunting and fishing licenses for reservation lands and waters. Some have their own wildlife conservation programs. The Ojibwas of the Red Lake Reservation in Minnesota created a sanctuary to protect bald eagles and other birds and animals. Several tribes, including the Standing Rock Sioux, have their own herds of buffalo. Sometimes game management is practiced as a source of tribal income; the Jicarilla Apaches, for example, have a large game park surrounded by a fence high and strong enough to keep deer and elk contained, where sportsmen from outside the reservation are allowed to hunt in a carefully supervised manner, and are charged rather substantial fees for the animals actually killed. Several tribes have guide services available for hunting parties. Sometimes jurisdictional disputes with the state have developed over use of these resources; the State of Wisconsin, for example, has tried to regulate the harvesting of wild rice in ways that Indians consider unfair.

The airplane has had far-reaching effects in more isolated areas where the terrain has kept the land inaccessible by road even today, as in Alaska. Hunters and fishermen, not to mention forest and mineral surveyors, can fly into the remotest lakes or inlets by airplanes equipped with pontoons. Laws against hunting from airplanes are difficult to enforce in this still relatively uncrowded region.

CHAPTER IX

Indian forests have also suffered widespread and destructive impacts up to the present time. Logging on private lands and in government-administered forests has almost always consisted of clear-cutting whole tracts, since that method is economically more advantageous than selectively taking out a few trees here and there. As a result, hillsides are denuded and subjected to erosion. Streams, polluted by pulp mill waste and choked with silt and logging debris, may dry up in summer, and are no longer available as salmon spawning grounds. Since salmon will ordinarily not go to another stream, the size of the run is reduced. Millions of logs, representing a small part of the waste involved, line every beach along the Northwest Coast and mills have long polluted the air. The redwoods in northern California are being cut rapidly; the only mature trees left in another decade or two will be inside state and national parks, and even there the problems of erosion from deforested watershed are severe. Meanwhile, logging and the activities associated with it have driven wildlife from one area to another, or destroyed its habitat. Some laudable efforts have been made in reforestation; the Haida, for example have been employed in collecting spruce and cedar seeds for use in tree farms, but the general picture is not good. Mining has also had unfortunate effects, polluting streams, increasing erosion, and forcing the removal of Indian villages.

Within the past decade, long-standing Indian demands for the return of lands from National Forests and National Parks to tribal ownership have met with success. The Yakimas of Washington received back 21,000 acres of National Forest land near Mount Adams which had once been excluded, in error, from their reservation. Perhaps the best known case is that in which Taos Pueblo was justly granted one of its sacred localities, Blue Lake, along with thousands of acres of surrounding National Forest land. The Taos tribe plans to keep the area inviolate, and use it only for religious purposes. The Havasupai tribe received the largest award of land in a bill passed by Congress in 1975, giving them many of their ancestral territories included in parts of Grand Canyon National Park and Monument, as well as some Forest Service and Bureau of Land Management areas. The Havasupai award contains a provision, accepted by the tribe, that most of the land will be kept in a near-wilderness state.

The Havasupai case produced a serious argument between those favoring the expansion of the Indian land base on the one hand, and environmentalists who opposed the removal of National Park and Forest lands from public ownership on the other. The dispute was most unfortunate, because Indians, with the strong ecological background of their traditional culture, should find natural allies among environmentalists who have come to appreciate the Indian contributions along these lines. But the issue must be faced. The conservation movement of the late nineteenth century and the first half of the twentieth had to fight tough political battles to save the natural resources of the United States from ruthless exploitation. Two different kinds of hard-won victories in these battles are represented in tangible form by the National Parks and National Forests, lands which were set aside for administration by government agencies on behalf of all the American people. National Parks were to be preserved in their natural state, with only minimal changes such as roads and campgrounds consistent with public access so that the magnificent scenery could be enjoyed. National Forests allowed for multiple use of resources by private citizens and companies, under government management which would ensure the conservation of renewable resources such as timber, grazing lands, and water. A few areas of the National Parks and Forests have been set aside as wilderness areas where permanent human residence and major signs of human presence are prohibited.

No one is seriously proposing that all these lands be returned to Indian tribes, although some claim might be made to all of them, since they were all once owned by Indian tribes in the past. In the Havasupai case, the land was returned even though the tribe had already accepted an award for some of the same land from the Indian Claims Commission. Each case of a claim for the return of public lands to Indian tribes must be considered on its own merits. But some principles should be clear. The American earth, particularly as represented in the public lands, is the patrimony of all the American people, incuding the American Indians. The administration of public lands must take into account the needs of Indian tribes as well as other Americans.

An example of a case where public lands were distributed to Indians not long ago was in Alaska, where they had never signed treaties, nor had they been assigned to reservations, with the exception of Metlakatla. The Alaska Native Settlement Act of 1971 permitted Indians to select land of their own, a total of forty million acres in all of Alaska, and each village will receive a portion of that, and also part of a $500 million payment in return for their aboriginal land rights in the rest of Alaska, as well as a two percent share of the oil revenues generated in Alaska.[49] The development of oil on the Alaskan North Slope and the completion of the Alaska oil pipeline does raise the very

CHAPTER IX

real possibility of pollution caused by oil spills from tankers traveling offshore along the Northwest Coast from Alaska southward, and consequent effects on fishing, wildlife, and other aspects of the regional ecology for coastal Indians as well as others.

Probably the most crucial threat to what remains of the ecological integrity of Indian lands today is energy development. It might be said that Smohalla and Wovoka had at least some ecological revenge after the dispossession of the Indians, when the plow broke the plains and loosed the dust bowl. But more may lie in the future. Many of the Indian Reservations are located above rich coal deposits that are close enough to the surface to be strip-mined. The nation demands energy, and the Indians need an economic base, so contract negotiations and litigation are in progress. What will happen to the ecosystems and the relationships between people and nature when the ground itself is moved away to get at the coal? What will the land be like, and what will have happened to the people, when the coal is gone? Environmental impact is accelerating today beyond anything seen before. Indian land also lies over deposits of gas, oil, and uranium at a time when national energy demands seem to override most other considerations. Strip mining is, in traditional Indian eyes, the ultimate desecration, the destruction of the land itself. Thousands of acres have been treated in this way, and the process has only begun. Reclamation of land after strip mining in the arid Southwest and the northern plains with a short growing season would have to take account of the fragility of the ecosystems and the danger of erosion on a monumental scale. It would involve great expense and in many places it is not being done.

Coal transportation through slurry pipelines carries millions of gallons of water away from a region where it is desperately needed. Huge power generation plants in Indian country have not observed the regulations against air pollution that apply in other areas; even larger plants are being planned, and the formerly clear air of the Southwest is visibly more turbid every year. The mesas and canyons are spanned by giant power transmission lines supplying energy to distant cities. Pueblo lands have become sites for White suburban and vacation home development. Indian people have been forced to move from the paths of mining and power projects, had their lives disrupted, and all too often have found jobs unavailable in the new projects where the immediate need is for highly trained and experienced workers. All these changes constitute an overwhelming denial of the traditional Indian view of nature, and tend either to prevent the traditional ways of interacting with nature, or corrupt them. Older ways that maintained the balance of nature sometimes became inappropriate in the new technological setting. Returning waste to the earth worked when almost everything discarded was biodegradable; it does not work with plastic containers and aluminum beer cans.[50] The principle of using every part of an animal, and throwing nothing away, is applied to automobiles today and, while generally it still works as a method of recycling, it has produced some unsightly heaps across the landscape.

The pressures for large-scale exploitation of Indian resources have never been stronger than they are today, and the attack on Indian environmental attitudes and practices through two hundred years of history has been all too successful, although not completely so. Within Indian tribes there are differences between those who favor full-scale economic development of all resources, and those who favor acting in accord with the best available ecological wisdom. Some tribal leaders want to guard their land, forests, water, and wildlife from developments that would pollute or destroy them, while others feel that Indian economic needs are so great that they should welcome industry of every kind and take their chances with environmental impact.[51] Thus, putting additional land under Indian tribal control will not guarantee its wise ecological use. It can be hoped, however, that Indians will determine to reassert the respect for life and the earth that their ancestors had, and demand that their lands be treated as a sacred trust from those venerable people and for their descendants. In order to do this, they will also need to incorporate the understandings of modern ecological scientists, and seek their advice. Native American Indian ecologists, educated in the best American universities, are especially needed to guide the management of Indian lands. It is still possible that Indian tribes today might make their lands ecological models for the world to admire and emulate. But it will take hard-won decisions at the tribal level, combined with federal support which enables the tribes to gain the necessary finances and scientific skills.

In several instances, Indian tribes are gaining control of the schools where their own children are educated. It is to be hoped that ample room will be found in the curriculum of these schools for environmental education which includes both the insights of Indian ecology and the best modern understandings of human interactions with nature.[52] Only the most traditional of Indians still retain their ecological religion intact, so that Indians today need scientific ecology and conservation as much as everyone else.

CHAPTER IX

It comes as no surprise that the impact of the alien culture has also distorted and fragmented the Indians' ecological attitudes and practices. What is surprising is that many of them have persisted until the present day. And they have persisted. The ways of life described above in the past tense are not entirely things of days gone by. Many Indians still feel the same way. When they go out to hunt by themselves, away from the eyes of Whites who do not understand, these few still keep the traditions and say the ancient prayers of thanks and atonement.

And many Indians have come to resent what they see as the destructive course of modern technology. The Lakota Sioux author, Vine Deloria, Jr., gave a statement of this feeling in strong terms:

> In recent years we have come to understand what progress is. It is the total replacement of nature by an artificial technology. Progress is the absolute destruction of the real world in favor of a technology that creates a comfortable way of life for a few fortunately situated people. Within our lifetime the difference between the Indian use of land and the white use of land will become crystal clear. The Indian lived with his land. *The white destroyed his land. He destroyed the planet earth.*[53]

In their opportunity to conserve their resources and use them wisely, Indians are a step ahead of the rest of American society, and for a brief time they have the opportunity to keep that lead. While most Americans have to create a new ecological attitude to meet the decisions of the immediate future, Indians have a heritage of ecological wisdom available for them in the ages-long experience of their ancestors on this continent. Most Indian lands are located in the West and Alaska, where a new cycle of large-scale resource development is just beginning, and where ground rules can still be designed to control development. A large proportion of the known energy reserves are located within Indian reservations. Widespread strip mining, in particular, must be controlled so as to avoid serious deterioration of the land base of Indian tribes. Indians must win additional safeguards that will protect their sacred earth and their community life.

As one of today's Indian poets, Winifred Fields Walters, has said, "life/ In tune with nature's balanced give and take" is a "timeless way, / So simple, so complex so nearly gone."[54] But it is also the only way which holds a lasting hope for survival and humanity's future.

A sense of place. This shrine marks a sacred location at Zuñi, New Mexico. The Zuñis give their own pueblo the name Itiwana, *meaning* Center of the Earth. *Photograph circa 1903.*
COURTESY MUSEUM OF THE AMERICAN INDIAN, HEYE FOUNDATION.

Atsina Indians move through the typical high plains environment. Photographer Edward S. Curtis, 1908.
COURTESY SMITHSONIAN INSTITUTION NATIONAL ANTHROPOLOGICAL ARCHIVES.

Destruction of ecosystems. A Navajo is shown with his herd of sheep, a goat, and his horse at a spring in overgrazed land near Keams Canyon, AZ. Photograph taken in 1897.

COURTESY MUSEUM OF THE AMERICAN INDIAN, HEYE FOUNDATION.

Buffalo, slaughtered by commercial hunters, are shown as they lie on the plains of Montana.
PHOTOGRAPH COURTESY MUSEUM OF THE AMERICAN INDIAN, HEYE FOUNDATION.

X
Indian Wisdom for Today

WHAT CAN WE LEARN from American Indians that will be of value in meeting today's ecological realities and making the decisions of the immediate future? People who are concerned with the deteriorating state of our environment have discovered the American Indians' feeling for the earth, and they recognize in the Indian a basic attitude toward the natural environment that contrasts refreshingly with the dominant attitudes of European-American society. As Wilbur Jacobs concluded, "After having studied a mass of evidence in the biological, physical, and social sciences, I am convinced that Indians were indeed conservators. They were America's first ecologists."[1] "It is painfully clear that the United States needs the Indians and their culture," wrote Stewart Udall, former Secretary of the Interior, "a culture that has a deep reverence for nature, and values the simple, the authentic, and the humane."[2]

The discovery of the American Indian by environmentally concerned Americans is not entirely new, however. It is quite possible that part of the strength of the conservation-ecology concern in America comes from the American Indians' presence here and their influence on our national life and thought. Henry Thoreau was fascinated with the Indians, and gave evidence of his study of their personality, culture, and harmony with nature in many pages of his books and journals. "The charm of the Indian to me," he wrote, "is that he stands free and unconstrained in nature, is her inhabitant and not her guest, and wears her easily and gracefully."[3] This is the life style he tried to emulate in *Walden*. In *The Maine Woods*, he speaks at length of his own experiences with Indians, especially his Penobscot guide, Joe Polis, a master of woods lore. John Wesley Powell, not only an explorer but also an influential mid-nineteenth century voice for conservation and reclamation, spent time among the Paiutes recording and preserving their traditions. He translated and published some of their poetry.[4] The

CHAPTER X

Bureau of American Ethnology, which he founded, did more than any other organization to disseminate accurate knowledge about the American Indians. John Muir, the great naturalist, defender of wilderness, and founder of the Sierra Club, often met and talked with Indians. He was adopted by the Stickeen band of the Tlingit tribe and given the name *Ancoutahan*. He discovered that when he talked to them of his interest in plants, glaciers, and other natural phenomena, he struck a chord of sympathy that the missionaries had never touched. They said they had never heard a White man speak as he did.[5] Listening to an Indian chant, he found that "falling boulders and rushing streams and wind tones caught from rock and tree were in it,"[6] all things that Muir himself loved. "To the Indian mind," he wrote, "all nature was instinct with deity. A Spirit was embodied in every mountain, stream, and waterfall."[7] He noticed their love of flowers and wondered at their knowledge of animals. Indians hold that animals have souls, he noted, and refuse to speak disrespectfully of them. One night on a canoe trip toward Chilkat, Alaska, Muir recorded a conversation in which the Indians imputed their own attitudes of conservation to wolves:

> I greatly enjoyed the Indians' camp-fire talk this evening on their ancient customs, how they were taught by their parents ere the whites came among them, their religion, ideas connected with the next world, the stars, plants, the behavior and language of animals under different circumstances, manner of getting a living, etc. When our talk was interrupted by the howling of a wolf on the opposite side of the strait, Kadachan puzzled the minister with the question, "Have wolves souls?" The Indians believe that they are wise creatures who know how to catch seals and salmon by swimming slyly upon them with their heads hidden in a mouthful of grass, hunt deer in company, and always bring forth their young at the same and most favorable time of the year. I inquire how it was that with enemies so wise and powerful the deer were not all killed. Kadachan replied that wolves knew better than to kill them all and thus cut off their most important food-supply.[8]

Muir said that the Indians' healthy supply of pure air and water is something that "civilized toilers might well envy."[9] His observation of Indians was far too acute to be sentimental, but he sympathized early with the view that "Indians, children of Nature, living on the natural products of the soil" should not be "robbed of their lands and pushed ruthlessly back into narrower and narrower limits by alien races who were cutting off their means of livelihood."[10] Thus Indian-inspired ideas were active in the minds of individuals like Thoreau, Powell, and Muir, who helped to form the conservationist philosophy in America.

Ecological patterns are always changing; the interactions observed today in an ecosystem are the result of other interactions that took place in the past. The subject of ecological studies is never static, and as long as there have been human beings, they have made changes in the natural environment which have in turn reacted to make changes in human cultures. As Robert F. Heizer expressed it, "ecology . . . is a dynamic situation, and its true significance can only be understood in the milieu of time and when analyzed by the historical method."[11] Ecologically speaking, the experience of the American Indians in the natural setting of this continent is the heritage of everyone who lives in North America today. As Vine Deloria says, "In seeking the religious reality behind the American Indian tribal existence, Americans are in fact attempting to come to grips with the land that produced the Indian tribal culture and their vision of community."[12]

One of the inescapable facts which emerge when we contrast the Indian past with the present is that the American Indians' cultural patterns, based on careful hunting and agriculture carried on according to spiritual perceptions of nature, actually preserved the earth and life on the earth. Since the period of colonization, wasteful destruction of the earth has accelerated. But we cannot convincingly argue that society today should return to Indian ways of life in their entirety; such atavism might be possible for small groups in isolated areas, but not for very many. Hunting and subsistence agriculture could support only a small fraction of our present population. Change is inevitable in all ecological systems, and change is especially rapid in North America today. The question, then, is the direction of change. Shall we continue to move toward ever more destructive use of natural resources, thus making necessary a harsh reckoning with nature and unwelcome constraints on our ways of life? Or shall we direct change as much as possible in the direction of harmony between human beings and the natural environment; toward a state in which we can both use and save, in which we will act with forbearance and nature will provide a sustained yield of renewable and recyclable resources? If we choose the second alternative, we can gain much by studying our American Indian heritage and seeking modern applications of the wisdom we find there.

CHAPTER X

How can we get in touch with the careful traditions in Indian cultures that took care of this land for centuries? Indians, of course, can do this by turning to their own tribal storehouses of wisdom. Non-Indians must decide that they want to seek those values for themselves. Indians long ago gave up trying to teach non-Indians the right ways to act toward the natural environment, because for many decades, very few non-Indians would listen. As Chief Tenaya of the Awanichis said to Major Savage, it was useless for Indians to try to tell anything to those "who know more than all the Indians."[13] Unfortunately for those who would like to learn from Indians, Tenaya's attitude is even stronger among Indians today. But it is still possible for individuals to find out what traditional Indian ecological values really are. There are many excellent studies written by historians, anthropologists, and Indians themselves, though one must discover how to separate the kernels from the husks. And sharing as an Indian way has not disappeared, even sometimes when it means sharing treasured insights with non-Indians who are able to receive them.

Certainly people today need to learn to see points of view other than those of their own cultural perspectives. One excellent way to do this is to listen to what Indians have been saying through the centuries about life and the earth; to try to view the world through Indian eyes as much as possible for a time. Indian conceptions of the universe and nature must be examined seriously, as valid ways of relating to the world, and not as superstitions, primitive, or unevolved. The American Indian experience was an integral way of life, an irreplaceable part of the total experience of mankind.

Perhaps the most important insight which can be gained from the Indian heritage is reverence for the earth and life. Indians did not advance this as a philosophical concept, but developed it by living with nature, depending on its cycles and interacting with the other forms of life. Indian respect for animals was based on observation of their ways within unhindered ecosystems. It springs not from sentiment, but from ethnoscience. Indians had a sense of reciprocity with life, of spiritual resonance with the natural environment, because the biosphere is truly alive and does interact with human beings in ways of its own. Viewed in this way, animals, plants, earth, air, and water have value that is intrinsic, not merely instrumental. In our way of taking nature and using it by turning it into a commodity, we have missed this value. In fact, industrial society has tended to treat human beings, as well as other forms of life, as "things" having merely instrumental value.

But the Indian heritage looks at the world as a whole, and sees human beings as part of nature. Sometimes one will hear it maintained as a point in argument that if mankind is part of nature, one need not worry about the impacts of culture on nature, since everything that people do must be "natural" by definition. But there are many things we can do. We can create, preserve, or destroy. We must choose among courses of action. And we cannot avoid the responsibility which results from our power over the earth and other forms of life. We do not have to damage nature, and therefore ourselves, just because we are capable of doing so.

It is important for us to learn from nature as the early Indians did, to keep an ear to the earth, and regain our perspective by frequent experiences with the non-artificial world, with animals and wild land. The ecological sciences can provide us with accurate knowledge that can be the basis for practices that sustain natural processes. Indians were always aware of the feedbacks provided by nature in response to their own actions; people today are not always so aware, because they are separated from the natural consequences of their actions by distance and artificial barriers. A washing machine in Los Angeles, for example, may be using energy generated by a power plant that fills the New Mexico air with smoke while it burns coal excavated from an Arizona strip mine occupying land where traditional Navajo Indians lived until recently, and which they call sacred. The sense of wholeness, of reciprocity with nature, needs to be restored. When early Indians took something from nature, they felt it imperative to give something back. What can we return to Mother Earth for the energy we are taking from her?

The traditional Indian view valued people, the interrelated social group living in harmony with nature in a particular landscape. Their sense of time was bound up with this concept of a people living in a land; to them, the unborn generations had a reality and a claim on the earth equal to their own. Historical continuity and human community were two aspects of the same thing, and both were expressed in the great tribal ceremonies. Can a sense of reverence for place be found again? Is it possible that in this time when community celebrations seem to have lost their cohesion, that a commitment to the localities in which we live might be expressed in revived Arbor Days or Earth Days? The decision to set aside and preserve the wildest and most beautiful parts of our continent as national parks and wilderness areas certainly deserves to be continued as recognition of the integrity of

CHAPTER X

special places. Even the building of houses that blend with the environment and climate, as do the tepee or Hopi pueblo; heated by the sun instead of irreplaceable fuels, might express the same intent. We have the knowledge and skills to move forward to a life more in harmony with nature, if we determine to do so.

Earlier in this chapter, it was said that only a very few people could manage to live a life today anything like that lived by the early Indians. There are those who would deny that such a life would even be desirable; after all, Mother Nature can be a harsh mistress, and comfort is certainly preferable to the ever-present possibility of thirst, hunger, and cold or heat. But as John Muir said about the Indians of the Grand Canyon, "The cañon Indians I have met here seem to be living much as did their ancestors. . . . They are able, erect men, with commanding eyes, which nothing that they wish to see can escape. They are never in a hurry. . . . Evidently their lives are not bitter." It was not a terrible existence for the American Indians in early times. The evidence of this comes from their own descriptions. They lived with the freedom and constraints of nature, and found life good. The Crow chief, Arapooish, stated it well:

> The Crow country is exactly in the right place. It has snowy mountains and sunny plains; all kinds of climates and good things for every season. When the summer heat scorches the prairies, you can draw up under the mountains, where the air is sweet and cool, and grass fresh and the bright streams come tumbling out of the snowbanks. There you can hunt the elk, the deer and the antelope, when their skins are fit for dressing; there you will find plenty of white bears and mountain sheep. In the autumn, when your horses are fat and strong from the mountain pastures, you can go down on the plains and hunt the buffalo, or trap beaver in the streams. And when winter comes on, you can take shelter in the woody bottoms along the rivers; there you will find buffalo meat for yourselves, and cottonwood bark for your horses; or you may winter in the Wind River Valley where there is salt weed in abundance. The Crow country is exactly in the right place. Everything good is to be found there. There is no country like the Crow country.[14]

Or as the Sioux, Ohiyesa (Charles Eastman) remarked in describing his orphaned childhood in a traditional Plains Indian band, "The freest life in the world . . . This life was mine. Every day there was a real hunt. There was real game. Occasionally there was a medicine dance away off in the woods. . ."[15]

Change is inevitable, of course, but the kind of change we will have can to some extent be chosen. It is human nature to use the natural environment, and therefore to change it. But this can be done in a way that preserves natural cycles and therefore preserves the human beings who depend on them, or it can be done in a way that disrupts natural cycles and deteriorates both the environment and human life. Indians found one way of doing the former; society today seems to be doing the latter. The value of Indian environmental perspectives may lie not in advising a return to earlier ways of subsistence, but in helping to develop a new style of life that incorporates care and reverence for nature and understands the limits that must be placed on human actions affecting the natural environment within the context of present knowledge and capabilities. One recalls the Navajo chant once more:

> May I walk in beauty of abundant rainshowers,
> May I walk in beauty of abundant vegetation,
> May I walk in beauty;
> With beauty before me, I walk,
> With beauty behind me, I walk,
> With beauty below me, I walk,
> With beauty above me, I walk.
> It is finished in beauty.[16]

One cannot listen to such a song without feeling that the people who sang it knew that their way was blessed with beauty. There were harsh realities in Indian life, but they were comprehended through the Indians' sense of being in harmony with a surrounding world of spiritual power. Today's life has more physical comforts, but it is not without its own harsh realities. Do we know a path where we, too, can walk in beauty?

NOTES

The poem on the dedication page is composed of selections from "The Hako: A Pawnee Ceremony," by Alice C. Fletcher, *Twenty-second Annual Report of the Bureau of American Ethnology*, 1900-01 (Washington: Government Printing Office, 1904), part 2, pp. 124-25.

Chapter 1

1. Melvin R. Gilmore, "Indians and Conservation of Native Life," *Torreya* 27 (November-December 1927): 98.
2. Luther Standing Bear, *Land of the Spotted Eagle* (Lincoln: University of Nebraska Press, 1978), p. 38.
3. There is full discussion of the idea that Indians didn't "use" the land in William T. Hagan, "Justifying Dispossession of the Indian: The Land Utilization Argument," in *American Indian Environments: Ecological Issues in Native American History*, ed. Christopher Vecsey and Robert W. Venables (Syracuse, New York: Syracuse University Press, 1980), pp. 65-80.
4. One of the most enlightening examinations of the opinions of both Indians and non-Indians supporting and opposing the idea that Indians were primal environmentalists is by Christopher Vecsey, "American Indian Environmental Religions," in *American Indian Environments*, ed. Vecsey and Venables, pp. 1-37.
5. Popovi Da, "Indian Values," *The Living Wilderness* 34 (Spring 1970): 25.

Chapter 2

1. Hal Borland, *When the Legends Die* (New York: Bantam Books, 1964), p. 18.
2. Frances Densmore, *Pawnee Music*. Bureau of American Ethnology, Bulletin No. 93 (Washington: Government Printing Office, 1929), p. 49.
3. Frances Densmore, *Teton Sioux Music*. Bureau of American Ethnology, Bulletin No. 61 (Washington: Government Printing Office, 1918), pp. 99-100.
4. Popovi Da, "Indian Values," *The Living Wilderness* 34 (Spring 1970): 26.
5. H.A. Smith, "Early Reminiscences, Nr. 10: Scraps from a Diary – Chief Seattle," etc., *Seattle Sunday Star*, October 29, 1887.
6. Ruth Murray Underhill, *Papago Indian Religion* (New York: Columbia University Press, 1946), pp. 33, 98; Willard W. Hill, *The Agricultural and Hunting Methods of the Navajo Indians* (Yale University Publications in Anthropology, No. 18, New Haven, 1938), p. 72.
7. John Collier, *On the Gleaming Way: Navajos, Eastern Pueblos, Zuñis, Hopis, and Their Land: and Their Meanings to the World* (Reprint. Denver: Sage Books, 1962), p. 132.
8. Underhill, *Papago Religion*, p. 238. Underhill's translation has "shaman" instead of "magician."
9. Ruth L. Bunzel, "Zuñi Ritual Poetry," *Forty-seventh Annual Report of the Bureau of American Ethnology*, 1929-30 (Washington: Government Printing Office, 1932), pp. 643-44.

Notes

10. Washington Matthews, "Navajo Legends," *American Folklore Society Memoirs* 5(1879): 273-75.

11. Black Elk, *The Sacred Pipe*, ed. Joseph Epes Brown (New York: Penguin Books, 1973), pp. 13-14.

12. Ibid., p. 7.

13. Elsie Clews Parsons, *Pueblo Indian Religion* (Chicago: University of Chicago Press, 1939), p. 198.

14. Elsie Clews Parsons, *Isleta Paintings*, ed. by Esther S. Goldfrank. Bureau of American Ethnology, Bulletin No. 181 (Washington: Government Printing Office, 1962), P. 78.

15. Edward F. Castetter and Morris Edward Opler, *The Ethnobiology of the Chiricahua and Mescalero Apache*. University of New Mexico Bulletin No. 297; Biological Series, vol. 4, no. 5, part 3 (Albuquerque: University of New Mexico Press, 1936), p. 16.

16. Joseph Epes Brown, "The Roots of Renewal," in *Seeing With a Native Eye: Essays on Native American Religion*, ed. Walter Holden Capps (New York: Harper & Row, 1976), p. 32.

17. Underhill, *Papago Religion*, pp. 22-23.

18. Parsons, *Pueblo Religion*, p. 207.

19. Black Elk, *The Sacred Pipe*, p. 105.

20. Ruth L. Bunzel, "Introduction to Zuñi Ceremonialism," *Forty-seventh Annual Report of the Bureau of American Ethnology*, 1929-30 (Washington, Government Printing Office, 1932), p. 488.

21. E. Adamson Hoebel, *The Cheyennes: Indians of the Great Plains* (New York: Holt, Rinehart and Winston, 1960), pp. 84-85.

22. Black Elk, *Sacred Pipe*, p. xx.

23. Ake Hultkrantz heard the Plains Cree Indian Stan Cuthand say, "The Supreme Being is in everything; he is in all of nature, in man and all the animals." *Belief and Worship in Native North America*, ed. Christopher Vecsey (Syracuse, New York: Syracuse University Press, 1981, p. 126.

24. Thomas F. McIlwraith, *The Bella Coola Indians*, 2 vols. (Toronto: University of Toronto Press, 1948), 1:32, 345.

25. John R. Swanton, *Contributions to the Ethnology of the Haida*. American Museum of Natural History, Memoirs, vol 8, part 1 (New York, 1909), p. 13.

26. Black Elk, *Sacred Pipe*, p. 14.

27. George S. Snyderman, "Concepts of Land Ownership Among the Iroquois and Their Neighbors," in *Symposium on Local Diversity in Iroquois Culture*, ed. by William N. Fenton. Smithsonian Institution, Bureau of American Ethnology, Bulletin 149 (1952); 16.

28. Ethelou Yazzie, ed., *Navajo History*, vol. 1 (Many Farms, Arizona: Navajo Community College Press, 1971), pp. 21, 23.

29. Frances Densmore, *Mandan and Hidatsa Music*. Bureau of American Ethnology, Bulletin No. 80 (Washington: Government Printing Office, 1923), p. 50; Densmore, *Teton Sioux Music*, p. 357.

30. George E. Hyde, *Life of George Bent, Written from His Letters* (Norman: University of Oklahoma Press, 1968), p. 155.

31. Black Elk, *Sacred Pipe*, pp. 12-13.

32. Franz Boas, *The Religion of the Kwakiutl Indians*, Part II: Translations (New York: Columbia University Press, 1930), p. 177.

Notes

33. McIlwraith, *Bella Coola*, 1:92.

34. Frances Densmore, *Nootka and Quileute Music*. Bureau of American Ethnology, Bulletin No. 124 (Washington: Government Printing Office, 1939), p. 282.

Chapter 3

1. Franz Boas, *The Religion of the Kwakiutl Indians*, 2 vols. (New York: Columbia University Press, 1930), 2: 191.

2. Luther Standing Bear, *Land of the Spotted Eagle* (Lincoln: University of Nebraska Press, 1978), p. 193.

3. Gladys A. Reichard, *Navajo Religion* (1950. Reprint. Princeton: Princeton University Press, 1963), p. 287.

4. Ake Hultkrantz, *Belief and Worship in Native North America*, ed. Christopher Vecsey (Syracuse, New York: Syracuse University Press), pp. 149-51.

5. Frank G. Speck, "Aboriginal Conservators," *Bird Lore (Audubon Magazine)* 40 (1939): 260. Italics are Speck's.

6. Irving Goldman, *The Mouth of Heaven; An Introduction to Kwakiutl Religious Thought* (New York: John Wiley and Sons, 1975), p. 53; cf. Franz Boas, *Contributions to the Ethnology of the Kwakiutl*. Columbia University, Contributions to Anthropology, 3 (New York, 1925), p. 357.

7. William Christie MacLeod, "Conservation Among Primitive Hunting Peoples," *Scientific Monthly* 43 (July-December 1936): 563.

8. Boas, *Religion of the Kwakiutl*, 2: 196-97.

9. Ibid., 2: 193.

10. Philip Drucker, *Culture Element Distributions: XXVI: Northwest Coast*. University of California, Anthropological Records, vol. 9, no. 3 (Berkeley, 1950), pp. 286-87; cf. pp. 222-23.

11. John R. Swanton, "The Tlingit Indians," *Twenty-sixth Annual Report of the Bureau of American Ethnology*, 1904-05 (Washington: Government Printing Offiice, 1908), p. 455.

12. John Witthoft, "The American Indian: Hunter," *Pennsylvania Game News* 29 (February 1953): 9.

13. Swanton, "Tlingit Indians," p. 455.

14. Francis La Flesche, "The Osage Tribe: Rite of the Wa-zo'-be," *Forty-fifth Annual Report of the Bureau of American Ethnology*, 1927-28 (Washington: Government Printing Offiice, 1930), p. 642.

15. Ruth Murray Underhill, *Papago Indian Religion* (New York: Columbia University Press, 1946), p. 16.

16. Willard W. Hill, *The Agricultural and Hunting Methods of the Navajo Indians* (Yale University Publications in Anthropology, No. 18, New Haven, 1938), p. 128.

17. Hamilton A. Tyler, *Pueblo Animals and Myths* (Norman: University of Oklahoma Press, 1975), p. 67; Reichard, *Navajo Religion*, p. 33.

18. Franz Boas, "Tsimshian Mythology," based on texts recorded by Henry W. Tate, *Thirty-first Annual Report of the Bureau of American Ethnology*, 1909-10 (Washington, Government Printing Office, 1916), p. 449.

19. Philip Drucker, *Cultures of the North Pacific Coast* (San Francisco: Chandler, 1965), pp. 133-38.

Notes

20. On this point, see the discussion in Ake Hultkrantz, "The Owner of the Animals in the Religion of the North American Indians," in *Belief and Worship in Native North America*, ed. Christopher Vecsey (Syracuse, New York: Syracuse University Press, 1981), pp. 135-146.
21. Frank G. Speck, "Savage Savers," *Frontiers* 4 (October 1939): 24.
22. Hamilton A. Tyler, *Pueblo Gods and Myths* (Norman: University of Oklahoma Press, 1964), p. 256.
23. Charles Hudson, "Cherokee Concept of Natural Balance," *Indian Historian* 3 (Fall 1970): 53.
24. Carl N. Gorman, "Navajo Vision of Earth and Man," *Indian Historian* 6 (Winter 1973): 21.
25. Fred Fertig, "Child of Nature: The American Indian as an Ecologist," *Sierra Club Bulletin* 55 (August 1970): 6.
26. Hill, *Agricultural and Hunting Methods of the Navajo*, p. 145.
27. Tyler, *Pueblo Animals and Myths*, p. 193.
28. Ernest Beaglehole, *Hopi Hunting and Hunting Ritual*. Yale University Publications in Anthropology, No. 4 (New Haven, 1936), p. 11.
29. Tyler, *Pueblo Animals and Myths*, pp. 36, 142.
30. Boas, "Tsimshian Mythology," p. 449.
31. Alvina Quam, trans., *The Zuñis: Self-Portrayals* (Albuquerque: University of New Mexico Press, 1972), p. 196.
32. Robert F. Heizer, "Primitive Man as an Ecologic Factor," *Kroeber Anthropological Society Papers* 13 (Berkeley, California: Fall 1955): 8.
33. Underhill, *Papago Religion*, pp. 245-46.
34. Frank G. Speck, "The Indians and Game Preservation," *The Red Man* 6 (September 1913): 23.
35. Speck, "Savage Savers," 26.
36. John R. Swanton, *Source Materials for the Social and Ceremonial Life of the Choctaw Indians*. Smithsonian Institution, Bureau of American Ethnology, Bulletin 103 (1931): 54, 101.
37. These quotations, and the one which follows, are from Frank Gilbert Roe, *The North American Buffalo: A Critical Study of the Species in its Wild State*, 2d ed. (Toronto: University of Toronto Press, 1970), p. 655.
38. Frank Gilbert Roe, *The Indian and the Horse* (Norman: University of Oklahoma Press, 1955), p. 357.
39. John C. Ewers, *The Blackfeet: Raiders on the Northwestern Plains* (Norman: University of Oklahoma Press, 1958), p. 11.
40. John C. Ewers, *Indian Life on the Upper Missouri* (Norman: University of Oklahoma Press, 1968), pp. 134-41.
41. Roe, *North American Buffalo*, p. 636.
42. Ewers, *Indian Life on the Upper Missouri*, pp. 166-67.
43. Joe Ben Wheat, "A Paleo-Indian Bison Kill," *Scientific American* 216 (January 1967): 44-52.
44. George C. Frison, ed., *The Casper Site: A Hell Gap Bison Kill on the High Plains* (New York: Academic Press, 1974), p. 104.
45. Roe, *North American Buffalo*, p. 655.
46. Parsons, *Pueblo Religion*, p. 718; Reichard, *Navajo Religion*, p. 81; Keith H. Basso, *The Cibecue Apache* (New York: Holt, Rinehart and Winston, 1970), p. 34.

Notes

47. Roe, *Indian and the Horse*, p. 376.
48. Black Elk, ed. by John G. Neihardt, *Black Elk Speaks* (1932. Reprint. Lincoln: University of Nebraska Press, 1961).
49. Roe, *Indian and the Horse*, p. 261; see also Gilbert Livingstone Wilson, *The Horse and the Dog in Hidatsa Culture*. American Museum of Natural History, Anthropological Papers, 15, part 2 (New York, 1924), pp. 151-56.
50. Roe, *Indian and the Horse*, pp. 252, 263; see also Ewers, *Horse in Blackfoot Culture*, pp. 46-47, 322.
51. Roe, *Indian and the Horse*, p. 265.
52. Wilson, *Horse and Dog in Hidatsa Culture*, p. 163.
53. James F. Downs, *The Navajo* (New York: Holt, Rinehart and Winston, 1972), p. 92.
54. Robert F. Heizer, "Primitive Man as an Ecologic Factor," *The Kroeber Anthropological Society Papers* 13 (Fall 1955): 7.
55. Erna Gunther, *Klallam Ethnography*. University of Washington Publications in Anthropology, vol. 1, no. 5 (Seattle: University of Washington Press, 1927), pp. 203-04; Drucker, *Cultures of the North Pacific Coast*, p. 95.
56. Boas, "Ethnology of the Kwakiutl," pp. 611-12.
57. McIlwraith, *Bella Coola*, 1: 74.
58. Wayne Suttles, "Affinal Ties, Subsistence, and Prestige among the Coast Salish," *American Anthropologist* 62 (1960): 302; Andrew P. Vayda, "A Re-examination of Northwest Coast Economic Systems," *Transactions of the New York Academy of Sciences*, series 2, 23 (1961): 619; Stuart Piddocke, "The Potlatch System of the Southern Kwakiutl: A New Perspective," *Southwestern Journal of Anthropology* 21 (1965): 244-45.
59. Charles Eastman, *The Soul of the Indian* (Boston: Houghton Mifflin Company, 1911; reprinted, New York: Johnson Reprint Corporation, 1971), pp. 9-10.

Chapter 4

1. Quoted in T. C. McLuhan, ed., *Touch the Earth* (New York: Promontory Press, c. 1971), p. 23.
2. Reichard, *Navajo Religion*, p. 22.
3. Leland Wyman and S. K. Harris, *Navajo Indian Medical Ethnobotany*. University of New Mexico Bulletin No. 366; Anthropological Series, vol. 3, no. 5 (Albuquerque: University of New Mexico Press, 1941), p. 7.
4. Ibid.
5. Ruth Murray Underhill, *Red Man's Religion; Beliefs and Practices of the Indians North of Mexico* (Chicago: University of Chicago Press, 1965), p. 111.
6. Collier, *On the Gleaming Way*, p. 45.
7. Castetter and Opler, *Ethnobiology of the Apache*, p. 17.
8. Frances Densmore, "Uses of Plants by the Chippewa Indians," *Forty-fourth Annual Report of the Bureau of American Ethnology*, 1926-27 (Washington: Government Printing Office, 1928), p. 314.
9. McIlwraith, *Bella Coola*, 1: 91-92.
10. Parsons, *Pueblo Religion*, p. 195; Underhill, *Papago Religion*, p. 16.
11. Quam, *Zuñis*, pp. 205-06.
12. Underhill, *Red Man's Religion*, p. 116.
13. Tyler, *Pueblo Gods and Myths*, pp. 251-52.

Notes

14. Edward S. Curtis, *The North American Indian*, vol. 10 (Cambridge, Massachusetts: The University Press, 1915), pp. 11-12.
15. Boas, "Ethnology of the Kwakiutl," pp. 616-17.
16. Ibid., p. 619.
17. Roe, *North American Buffalo*, p. 635.
18. Peter Kalm, *Peter Kalm's Travels to North America*, translated and edited by A. B. Benson (New York: Wilson-Erickson, 1937). Quoted in Gordon M. Day, "The Indian as an Ecological Factor in the Northeastern Forest," *Ecology* 34 (1953): 339.

Chapter 5

1. Reginald Laubin and Gladys Laubin, *The Indian Tipi: Its History, Construction and Use* (Norman: University of Oklahoma Press, 1957), p. 105.
2. See Joseph Epes Brown, "The Roots of Renewal," in *Seeing With A Native Eye: Essays on Native American Religion*, ed. Walter Holden Capps (New York: Harper & Row, 1976), pp. 29-30.
3. Yazzie, *Navajo History*, pp. 59-62.
4. N. Scott Momaday, "I Am Alive," in *The World of the American Indian* (Washington: National Geographic Society, 1974), p. 14.
5. Underhill, *Red Man's Religion*, p. 206; see also Alfonso Ortiz, "Ritual Drama and the Pueblo World View," in *New Perspectives on the Pueblos*, ed. by Alfonso Ortiz (Albuquerque: University of New Mexico Press, 1972), pp. 142-43; and Edward P. Dozier, *The Pueblo Indians of North America* (New York: Holt, Rinehart and Winston, 1970), p. 208.
6. Collier, *On the Gleaming Way*, p. 124.
7. Jimmie Durham, "Who Has the Right to Destroy a Species?" *Los Angeles Times*, July 2, 1978. Quoted in Edwin P. Pister, "Endangered Species: Costs and Benefits," *Environmental Ethics* 1 (Winter, 1979): 347.
8. Gorman, "Navajo Vision of Earth and Man," p. 19.
9. Hill, *Agricultural and Hunting Methods of the Navajo*, p. 21.
10. H.A. Smith, "Early Reminiscences, Nr. 10: Scraps from a Diary – Chief Seattle," etc., *Seattle Sunday Star*, October 29, 1887.
11. Snyderman, "Land Ownership among the Iroquois," 23.
12. Virginia Irving Armstrong, compiler, *I Have Spoken*, (Chicago: Swallow Press, 1971), p. 115.
13. Stewart Lee Udall, *The Quiet Crisis* (New York: Holt, Rinehart & Winston, 1963), p. 20.
14. Collier, *On the Gleaming Way*, p. 20.
15. Edward Sapir and Harry Hoijer, *Navaho Texts* (Iowa City: University of Iowa Press, 1942), p. 359.
16. Nabokov, *Native American Testimony*, pp. 107-08.

Chapter 6

1. Castetter and Bell, *Pima and Papago Agriculture*, p. 173.
2. Hill, *Agricultural and Hunting Methods of the Navajo*, p. 53.
3. Witthoft, "Hunter," 29 (February 1953): 15.
4. G. Browne Goode, "The Use of Agricultural Fertilizers by the American Indians and the Early English Colonists," *The American Naturalist* 14 (July 1880): 475. Heizer, "Primitive Man as an Ecologic Factor," 14-15.
5. Black Elk, *The Sacred Pipe*, p. 107.

Notes

6. George F. Will and George E. Hyde, *Corn Among the Indians of the Upper Missouri* (Lincoln: University of Nebraska Press, 1917), pp. 260-67.

7. Richard I. Ford, "An Ecological Perspective on the Eastern Pueblos," in *New Perspectives on the Pueblos*, ed. by Alfonso Ortiz (Albuquerque: University of New Mexico Press, 1972), p. 7.

8. Edward F. Castetter and Willis H. Bell, *Pima and Papago Indian Agriculture* (Albuquerque: University of New Mexico Press, 1942), p. 56.

9. Herbert W. Dick, *Bat Cave*. The School of American Research, Monograph No. 27 (Santa Fe, 1965).

10. Kirk Bryan, "Flood-Water Farming," *Geographical Review* 19 (1929): 445.

11. Guy R. Stewart and Maurice Donnelly, "Soil and Water Economy in the Pueblo Southwest," *Scientific Monthly* 56 (1943): 31-44, 134-44.

12. Castetter and Bell, *Pima and Papago Agriculture*, p. 169.

13. George F. Carter, *Plant Geography and Culture History in the American Southwest*. Viking Fund Publications in Anthropology, No. 5 (New York, 1945), p. 109.

14. Guy R. Stewart, "Conservation in Pueblo Agriculture," *Scientific Monthly* 51 (1940): 332.

15. Parsons, *Pueblo Indian Religion*, p. 31.

16. Castetter and Bell, *Pima and Papago Agriculture*, p. 229.

17. Ruth Murray Underhill, *Red Man's America: A History of Indians in the United States* (Chicago: University of Chicago Press, 1953), p. 196,

18. Juan Sinyella, "Havasupai Traditions," ed. by J. Donald Hughes, *Southwest Folklore* 1 (Spring 1977): 35-52.

19. Hill, *Agricultural and Hunting Methods of the Navajo*, p. 52.

20. Underhill, *Red Man's Religion*, p. 241.

21. Parsons, *Pueblo Religion*, p. 171.

22. Underhill, *Red Man's Religion*, p. 245.

23. Ibid., p. 247. Underhill's translation has "drunk" instead of "soaked."

24. Matilda Coxe Stevenson, "The Sia," *Eleventh Annual Report of the Bureau of American Ethnology*, 1889-90 (Washington: Government Printing Office, 1894), p. 85.

25. Ruth Murray Underhill, *Ceremonial Patterns in the Greater Southwest*. American Ethnological Society Monographs 13 (1948): 26.

26. Stewart and Donnelly, "Soil and Water Economy," p. 144.

27. Stewart, "Conservation in Pueblo Agriculture," p. 337.

Chapter 7

1. This is a paraphrase of Ella E. Clark, *Indian Legends of the Pacific Northwest* (Berkeley: University of California Press, 1953), pp. 61-63.

2. Black Elk, ed. by Joseph Epes Brown, *The Sacred Pipe: Black Elk's Account of the Seven Rites of the Oglala Sioux* (1953. Reprint. Baltimore: Penguin Books, 1971), p. 72.

3. Eastman, *Soul of the Indian*, p. 15.

4. Densmore, *Teton Sioux Music*, p. 172.

5. Standing Bear, *Land of the Spotted Eagle*, p. 194.

6. George Clutesi, *Son of Raven, Son of Deer: Fables of the Tse-Shaht People* (Sidney, British Columbia: Gray's Publishing, 1967), p. 9.

7. Underhill, *Papago Religion*, p. 42.

Notes

8. Alice C. Fletcher and Francis La Flesche, "The Omaha Tribe," *Twenty-seventh Annual Report of the Bureau of American Ethnology*, 1905-06 (Washington: Government Printing Office, 1911), p. 600.
9. Black Elk, *The Sacred Pipe*, p. 46.
10. Densmore, *Mandan and Hidatsa Music*, p. 71.
11. Ibid., p. 88.
12. Densmore, *Pawnee Music*, p. 57.
13. Ibid., p. 43.
14. Densmore, *Teton Sioux Music*, p. 186.
15. Drucker, *Cultures of the North Pacific Coast*, pp. 174-75.
16. H. R. Hays, *Children of the Raven: The Seven Indian Nations of the Northwest Coast* (New York: McGraw-Hill, 1975), pp. 196-97.
17. Drucker, *Cultures of the North Pacific Coast*, p. 98.
18. Ruth M. Underhill, *Red Man's Religion* (Chicago: University of Chicago Press, 1965), p. 63.
19. Jackson Steward Lincoln, *The Dream in Primitive Cultures* (1935. Reprint. New York: Johnson Reprint Corporation, 1970).
20. Robert H. Lowie, *Indians of the Plains* (New York: McGraw-Hill, 1954), p. 155.
21. Underhill, *Papago Religion*, p. 135.
22. Byron Harvey, III, "An Overview of Pueblo Religion," in *New Perspectives on the Pueblos*, ed. by Alfonso Ortiz (Albuquerque: University of New Mexico Press, 1972), p. 207.
23. Alexander MacGregor Stephen, *Hopi Indians of Arizona*. Southwest Museum, Leaflet No. 14 (Los Angeles, 1940), p. 18.
24. Fred Fertig, "Child of Nature: The American Indian as an Ecologist," *Sierra Club Bulletin* 55 (August 1970): 6.
25. Densmore, *Nootka and Quileute Music*, pp. 333, 335, 95, 196.

Chapter 8

1. Hal Borland, When the Legends Die (New York: Bantam Books, 1964), p. 18.
2. Tyler, *Pueblo Animals and Myths*, p. 167.
3. Hudson, "Cherokee Concept of Natural Balance," p. 54.
4. Wilbur R. Jacobs, "The Tip of an Iceberg: Pre-Columbian Indian Demography and Some Implications for Revisionism," *William and Mary Quarterly* 31 (January 1974): 123-32.
5. Hoebel, *Cheyennes*, p. 84.
6. Robert F. Heizer, "Primitive Man as an Ecologic Factor," *The Kroeber Anthropological Society Papers* 13 (Fall 1955): 15. Wilbur R. Jacobs, "The Indian and the Frontier in American History: A Need for Revision," *Western Historical Quarterly* 4 (January 1973): 50. Virgil J. Vogel, *American Indian Medicine* (Norman: University of Oklahoma Press, 1970), pp. 227-31.
7. Roderick Nash, ed., *Environment and Americans: The Problem of Priorities* (New York: Holt, Rinehart & Winston, 1972), p. 75.
8. James A. Tuck, "The Iroquois Confederacy," *Scientific American* 224 (February 1971): 32-42.
9. Melvin L. Fowler, "A Pre-Columbian Urban Center on the Mississippi," *Scientific American* 233 (August 1975): 101.
10. Ibid.

11. Vincent Scully, *Pueblo: Mountain, Village, Dance* (New York: Viking Press, 1975), p. 9.

12. Ernest Beaglehole, *Notes on Hopi Economic Life*. Yale University Publications in Anthropology, No. 15 (New Haven, 1937), p. 58.

13. Douglas W. Schwartz, "Climate Change and Culture History in the Grand Canyon Region," *American Antiquity* 22 (1957): 372-77.

14. Kirk Bryan, "Pre-Columbian Agriculture in the Southwest, as Conditioned by Periods of Alluviation," *Annals of the Association of American Geographers* 31 (1941): 227.

15. Richard Benjamin Woodbury, "Climatic Changes and Prehistoric Agriculture in the Southwestern United States," *Annals of the New York Academy of Sciences* 95 (1961): 708.

16. Kirk Bryan, "Date of Channel Trench (Arroyo) Cutting in the Arid Southwest," *Science* 62 (1925): 340, 343-44.

17. Kirk Bryan, *The Geology of Chaco Canyon, New Mexico, in Relation to the Life and Remains of the Prehistoric Peoples of Pueblo Bonito*. Smithsonian Miscellaneous Collections, No. 122 (Washington, 1954), p. 51.

18. Patricia J. Rand, "Factors Related to the Distribution of Ponderosa and Pinyon Pines at Grand Canyon, Arizona," Ph.D. Dissertation, Duke University, Department of Botany, 1965.

19. Maitland Bradfield, *Changing Pattern of Hopi Agriculture* (London: Royal Anthropological Institute, 1971), p. 45.

20. J. C. Kelley, "Factors Involved in the Abandonment of Certain Peripheral Southwestern Settlements," *American Anthropologist* 54 (1952): 381-82.

21. Scully, *Pueblo*, p. 24.

Chapter 9

1. Dee Brown, *Bury My Heart at Wounded Knee* (New York: Bantam Books, c. 1970), p. 304.

2. Stewart L. Udall, *The Quiet Crisis* (New York: Holt, Rinehart and Winston, 1963), p. 22.

3. Sylvester M. Morey, ed., *Can the Red Man Help the White Man? Denver Conference with the Indian Elders, 1968* (New York: G. Church, 1970), p. 18.

4. Christopher Vecsey, "American Indian Environmental Religions," in *American Indian Environments: Ecological Issues in Native American History*, eds. Vecsey and Venables (Syracuse, New York: Syracuse University Press, 1980), p. 3.

5. American Friends Service Committee, *Uncommon Controversy: Fishing Rights of the Muckleshoot, Puyallup, and Nisqually Indians* (Seattle: University of Washington Press, 1970), p. 71.

6. Roe, *North American Buffalo*, p. 614. Italics in original.

7. Ewers, *Blackfeet*, pp. 22, 29.

8. James H. Howard, "Dakota Winter Counts as a Source of Plains History," *Anthropological Papers*, No. 61, Bureau of American Ethnology, Bulletin No. 173 (Washington: Government Printing Office, 1960), p. 371.

9. American Friends Service Committee, *Uncommon Controversy*.

10. *Ibid.*, p. 15.

11. *Ibid.*, p. 35.

Notes

12. H.A. Smith, "Early Reminiscences, Nr. 10: Scraps from a Diary – Chief Seattle," etc., *Seattle Sunday Star*, October 29, 1887.

13. J. Donald Hughes, "The De-racialization of Historical Atlases: A Modest Proposal," *The Indian Historian* 7 (Summer 1974): 55-56.

14. Roe, *North American Buffalo*, p. 811.

15. Richard Erdoes, *The Sun Dance People* (New York: Alfred A. Knopf, 1972), p. 171.

16. James Mooney, "The Ghost-Dance Religion and the Sioux Outbreak of 1890," *Fourteenth Annual Report of the Bureau of American Ethnology*, 1892-93 (Washington: Government Printing Office, 1896), p. 825.

17. Roe, *North American Buffalo*, p. 639.

18. James Mooney, "Calendar History of the Kiowa Indians," *Seventeenth Annual Report of the Bureau of American Ethnology*, 1895-96 (Washington: Government Printing Office, 1898), p. 161.

19. Roe, *North American Buffalo*, p. 187.

20. Ewers, *Blackfeet*, p. 32.

21. Hays, *Children of the Raven*, p. 43.

22. George A. Pettit, *The Quileute of La Push, 1775-1945*. University of California, Anthropological Records, vol. 14, no. 1 (Berkeley: University of California Press, 1950), p. 43.

23. Pettit, *Quileute*, pp. 44-45.

24. McIlwraith, *Bella Coola*, 1: 31.

25. Wilbur R. Jacobs, "Frontiersmen, Fur Traders, and Other Varmints, An Ecological Appraisal of the Frontier in American History," *American Historical Association Newsletter* (November 1970): 5-11, and "The Indian and the Frontier in American History: A Need for Revision," *Western Historical Quarterly* 4 (January 1973): 43-56.

26. S. Lyman Tyler, *A History of Indian Policy* (U.S. Department of the Interior, Bureau of Indian Affairs, 1973), p. 99.

27. Densmore, *Teton Sioux Music*, p. 173.

28. Drucker, *Cultures of the North Pacific Coast*, pp. 208-12.

29. Parsons, *Pueblo Religion*, p. 475.

30. Barre Toelken, "Seeing With a Native Eye: How Many Sheep Will It Hold?" in *Seeing With a Native Eye: Essays on Native American Religion*, ed. Walter Holden Capps (New York: Harper & Row, 1976), p. 14.

31. Mooney, "Ghost-Dance," pp. 723-24.

32. Ibid., p. 721.

33. Roe, *North American Buffalo*, p. 645.

34. Erdoes, *Sun Dance People*, p. 180.

35. David H. Miller, *Ghost Dance* (New York: Duell, Sloan and Pearce, 1959), p. 102.

36. Mooney, "Ghost-Dance," p. 1072.

37. Ibid., p. 1084.

38. Densmore, *Teton Sioux Music*, p. 173.

39. John Collier, *Indians of the Americas* (1947. Reprint. New York: New American Library, n. d.), p. 8.

40. Christopher Vecsey and Robert W. Venables, eds., *American Indian Environments: Ecological Issues in Native American History* (Syracuse, New York: Syracuse University Press, 1980), p. xix.

41. Henry W. Hough, *Development of Indian Resources* (Denver: World Press, 1967), p. 172.

Notes

42. Bryan, "Pre-Columbian Agriculture," pp. 219-42.
43. Collier, *On the Gleaming Way*, p. 62.
44. Hough, *Development of Indian Resources*, p. 165.
45. James E. Officer, "The American Indian and Federal Policy," in *The American Indian in Urban Society*, ed. by Jack O Waddell and O. Michael Watson (Boston: Little, Brown, 1971), pp. 45-47.
46. Edwin Pister, "Endangered Species: Costs and Benefits," *Environmental Ethics*, I, 4 (Winter 1979): 347-48.
47. American Friends Service Committee, *Uncommon Controversy*, pp. 64-65.
48. Ibid., pp. 141-42.
49. D'Arcy McNickle, *Native American Tribalism: Indian Survivals and Renewals* (London: Oxford University Press, 1973), p. 158.
50. The same point is made by Ake Hultkrantz, *Belief and Worship in Native North America*, ed. Christopher Vecsey (Syracuse, New York: Syracuse University Press, 1981), p. 124.
51. Tyler, *A History of Indian Policy*, p. 264.
52. Dave Warren, "Man and His Environment: The American Indian and the Natural World," in *Environmental Awareness for Indian Education*. U.S. Department of the Interior, Bureau of Indian Affairs, Curriculum Bulletin no. 8 (1970), pp. 5-11.
53. Vine Deloria, Jr., *We Talk, You Listen* (New York: Macmillan, 1970), p. 186.
54. Winifred Fields Walters, "Navajo Signs," in *Voices from Wah'kon-tah; Contemporary Poetry of Native Americans*, ed. by Robert K. Dodge and Joseph B. McCullough (New York: International Publishers, 1974), pp. 119-20.

Chapter 10

1. Wilbur R. Jacobs, "Indians as Ecologists and Other Environmental Themes in American Frontier History," in *American Indian Environments*, eds. Vecsey and Venables, p. 49.
2. Stewart L. Udall, "The Indians: First Americans, First Ecologists," in *Look to the Mountain Top*, ed. by Robert L. Iacopi, Bernard L. Fontana, and Charles Jones (San Jose, California: Gousha Publications, 1972), p. 6.
3. Henry Thoreau, *Writings*, Manuscript Edition, 7: 253, quoted in Roy Harvey Pearce, *The Savages of America* (Baltimore: The Johns Hopkins Press, 1965), p. 148.
4. John Wesley Powell, "Sketch of the Mythology of the North American Indians," *First Annual Report of the Bureau of American Ethnology*, 1879-80 (Washington, Government Printing Office, 1881), p. 23; see also the original, full text of *Exploration of the Colorado River of the West and Its Tributaries* (Washington, Government Printing Office, 1875).
5. John Muir, *The Writings of John Muir*, Manuscript Edition, 10 vols. (Boston: Houghton Mifflin, 1916-24), 3: 208-11.
6. Ibid., 10: 22.
7. John Muir, *John of the Mountains: The Unpublished Journals of John Muir*, ed. by Linnie Marsh Wolfe (Boston: Houghton Mifflin, 1938), p. 315.
8. Muir, *Writings*, 3: 151-52.
9. Ibid., 2: 206.
10. Ibid., 1: 174-75.

Notes

11. Robert F. Heizer, "Primitive Man as an Ecologic Factor," *Kroeber Anthropological Society Papers* 13 (Berkeley, California, Fall 1975): 2.

12. Vine Deloria, *God is Red* (New York: Grosset & Dunlap, 1973), p. 88.

13. Muir, *Writings*, 5: 257.

14. Stewart L. Udall, *The Quiet Crisis* (New York: Holt, Rinehart and Winston, 1963), pp. 17-18.

15. Charles A. Eastman, *Indian Boyhood* (New York: McClure, Phillips, 1902), p. 3.

16. Washington Matthews, "Navajo Legends," *American Folklore Society Memoirs* 5 (1897): 275, slightly altered.

SELECTED BIBLIOGRAPHY

This list contains only books and articles which study the relation of the American Indians to the natural environment, with a few added which are of exceptional value for other reasons in elucidating the subject of this book. To have listed all works used in preparation of this book would have produced a very long bibliography to little purpose. The reader is directed also to works listed in the notes.

Alexander, Hartley Burr. *The World's Rim: Great Mysteries of the North American Indians.* Lincoln: University of Nebraska Press, 1953.

Beaglehole, Ernest. *Hopi Hunting and Hunting Ritual.* New Haven: Yale University *Publications in Anthropology,* No. 4 (1936).

Benedict, Ruth F. "The Concept of the Guardian Spirit in North America," American Anthropological Association, *Memoirs,* No. 29 (1923).

————. "The Vision in Plains Culture," *American Anthropologist,* 24 (1922).

Bilsky, Lester J., ed. *Historical Ecology: Essays on Environment and Social Change.* Port Washington, N.Y.: National University Publications, Kennikat Press, 1980. Note particularly "Prehistoric Ecological Crises" by Michael P. Hoffman, pp. 33-42.

Black Elk, ed. John G. Neihardt. *Black Elk Speaks: Being the Life Story of a Holy Man of the Oglala Sioux.* Lincoln, Neb.,: University of Nebraska Press, (1932) 1961.

Black Elk, ed. Joseph Epes Brown. *The Sacred Pipe: Black Elk's Account of the Seven Rites of the Oglala Sioux.* Norman, Okla.: University of Oklahoma Press, 1953. (Reprint: Baltimore, Md.: Penguin Books, 1971).

Brown, Joseph Epes. *The Spiritual Legacy of the American Indian.* Lebanon, Pa.: Pendle Hill, 1964.

Capps, Walter Holden, ed. *Seeing with a Native Eye: Essays on Native American Religion.* New York: Harper & Row, 1976.

Carter, George F. "Ecology – geography – ethnobotany," *Scientific Monthly* 70 (1950), pp. 73-80.

Collier, John. *Indians of the Americas.* np, 1947. (Reprint: New York: New American Library, nd).

————. *On the Gleaming Way: Navajos, Eastern Pueblos, Zunis, Hopis, Apaches, and Their Land; and their Meanings to the World.* Denver: Sage Books, 1962.

Crosby, Alfred W., Jr. *The Columbian Exchange: Biological and Cultural Consequences of 1492.* Westport, Conn.: Greenwood Publishing Co., 1972.

Cushing, Frank Hamilton. "Zuni Breadstuff," *Indian Notes and Monographs,* 8 (1920), New York: Museum of the American Indian, Heye Foundation, Reprint, 1974.

————. "Zuni Fetishes," *Second Annual Report of the Bureau of Ethnology* to the Secretary of the Smithsonian Institution, 1880-1881, by J. W. Powell, Director. Washington, Government Printing Office, 1883, pp. 3-45.

Da, Popovi. "Indian Values," *The Living Wilderness,* vol. 34, no. 109 (Spring, 1970), pp. 25-26.

Davidson, D.S. "Family Hunting Territories in Northwestern North America," *Indian Notes and Monographs,* 46 (1928), New York: Museum of American Indian, Heye Foundation.

Day, Gordon M. "The Indian as an Ecological Factor in the Northeastern Forest," *Ecology,* vol. 34 (1953), pp. 329-46.

Bibliography

Deloria, Vine, Jr. *God is Red.* New York: Grosset & Dunlap, 1973.

Densmore, Frances. *The Belief of the Indian in a Connection Between Song and the Supernatural.* BAE, Bulletin 151, no. 37 (1953).

———. "Notes on the Indians' Belief in the Friendliness of Nature," *Southwestern Journal of Anthropology,* 4 (1948):94-97.

Downs, J. F. *Animal Husbandry in Navajo Society and Culture.* University of California, *Publications in Anthropology,* vol. 1 (1964).

———. "Domestication: An Examination of the Changing Relations Between Man and Animals," *Kroeber Anthropological Society Papers* 22 (1964): 18-67.

Dobyns, Henry F. and Robert C. Euler. *The Ghost Dance of 1889 among the Pai Indians of Northwestern Arizona.* Prescott, Arizona: Prescott College Press, 1967.

Drucker, Phillip. "A Karuk world-renewal ceremony at Panaminik," *University of California Publications in American Archaeology and Ethnology,* vol. 35, no. 3, 23-28. Berkeley: University of California Press, 1936.

——— and Robert Heizer. *To Make My Name Good.* Berkeley: University of California, 1967.

Eastman, Charles A. *Indian Boyhood.* New York: McClure, Phillips & Co., 1902.

———. *The Soul of the Indian.* New York: Houghton Mifflin, 1911.

Ewers, John C. *The Horse in Blackfoot Indian Culture, with Comparative Material from Other Western Tribes,* BAE, Bulletin No. 159 (1955).

Fertig, Fred. "Child of Nature: The American Indian as an Ecologist," *Sierra Club Bulletin,* vol. 55, no. 8 (August 1970), pp. 4-7.

Fewkes, Jesse Walter. "Sun Worship of the Hopi Indians," *Smithsonian Institution Annual Report for 1918.* Washington: Government Printing Office, 1920, pp. 493-526.

Frison, George C., ed. *The Casper Site: A Hell Gap Bison Kill on the High Plains.* New York: Academic Press, 1974.

Gilmore, Melvin R. "Indians and Conservation of Native Life," *Torreya,* vol. 27, no. 6 (Nov.-Dec., 1927), pp. 97-98.

Goode, G. Browne. "The Use of Agricultural Fertilizers by the American Indians and the Early English Colonials," *American Naturalist* 14 (1880), 473-79.

Gorman, Carl N. "Navajo Vision of Earth and Man," *Indian Historian,* vol. 6, no. 1 (Winter, 1973), pp. 19-22.

Gunther, Erna. "A Further Analysis of First Salmon Ceremony," *University of Washington Publications in Anthropology,* vol. III, no. 5. Seattle: University of Washington Press, 1928.

Hack, J. T. *The Changing Physical Environment of the Hopi Indians of Arizona.* Cambridge, Massachusetts: Peabody Museum of American Archaeology and Ethnology Papers, vol. 36.

Haeberlin, H. K. "The Idea of Fertilization in the Culture of the Pueblo Indians," American Anthropological Association, *Memoirs* 3 (1916), pp. 1-55.

Haile, Berard. *Starlore among the Navajo.* Santa Fe: Museum of Navajo Ceremonial Art, 1947.

Hallowell, A. I. "Bear Ceremonialism in the Northern Hemisphere," *American Anthropologist* 28 (1926), 1-175.

Heizer, Robert F. "Primitive Man as an Ecologic Factor," *The Kroeber Anthropological Society Papers,* no. 13 (Fall, 1955), pp. 1-31.

Bibliography

Highwater, Jamake. *The Primal Mind: Vision and Reality in Indian America.* New York: Harper & Row, 1981.

―――――. *Ritual of the Wind: North American Indian Ceremonies, Music, and Dances.* New York: Viking Press, 1977.

Hill, Willard W. "The Agricultural and Hunting Methods of the Navajo Indians," *Yale University Publications in Anthropology*, no. 18 (1938).

―――――. "Navajo Salt Gathering," *University of New Mexico Bulletin No. 349, Anthropology Series*, vol. 3, no. 4 (1940).

Hill, Willard W. and Dorothy W. Hill. "The Legend of the Navajo Eagle Catching Way," *New Mexico Anthropologist* 6, 7 (1943), 31-36.

Hudson, Charles. "Cherokee Concept of Natural Balance," *Indian Historian*, vol. 3, no. 4 (Fall, 1970), pp. 51-54.

Hughes, J. Donald. "Forest Indians: The Holy Occupation," *Environmental Review* 1 (1977):2-13.

Hultkrantz, Ake. "Attitudes to Animals in Shoshoni Indian Religion," *Studies in Comparative Religion* 4 (1970):70-79.

―――――. *Belief and Worship in Native North America.* Edited, with an introduction by Christopher Vecsey. Syracuse, N.Y.: Syracuse University Press, 1981.

―――――. "An Ecological Approach to Religion," *Ethnos* 31 (1966):131-50.

―――――. "Ecology of Religion: Its Scope and Methodology," *Studies in Methodology*, ed. by Lauri Honko. The Hague, Mouton, 1979, pp. 221-36.

―――――. "The Indians and the Wonders of Yellowstone: A Study of the Interrelations of Religion, Nature, and Culture," *Ethnos* 19 (1954):34-68.

―――――. "The Owners of the Animals in the Religion of the North American Indians," *The Supernatural Owners of Nature*, ed. by Ake Hultkranz. Stockholm: Acta Universitatis Stockholmiensis, Stockholm Studies in Comparative Religion 1 (1961):53-64.

―――――. "Religion and Ecology among the Great Basin Indians," *The Realm of the Extra-Human, Ideas, and Actions*, ed. Agehananda Bharati. The Hague: Mouton, 1976, pp. 137-50.

Jacobs, Wilbur R. *Dispossessing the American Indian: Indians and Whites on the Colonial Frontier.* New York: Scribner's, 1972.

―――――. "Frontiersmen, Fur Traders, and Other Varmints, An Ecological Appraisal of the Frontier in American History," *American Historical Association Newsletter* (Nov. 1970), pp. 5-11.

―――――. "The Indian and the Frontier in American History — a Need for Revision," *Western Historical Quarterly*, vol. 4, no. 1 (January 1973), 43-56.

Kelley, J. C. "Factor involved in the abandonment of Certain Peripheral Southwestern Settlements," *American Anthropologist* 54 (1952), 356-87.

Kidwell, Clara Sue. "Science and Ethnoscience," *The Indian Historian*, vol. 6, no. 4 (Fall, 1973), pp. 43-54.

Kroeber, Alfred Louis, and E. W. Gifford. *World Renewal: A Cult System of Native Northwest California.* Berkeley: University of California, Anthropological Records 13 (1), 1949.

Kurath, Gertrude Prokosh. "Effects of Environment on Cherokee-Iroquois Ceremonialism, Music, and Dance," *Bureau of American Ethnology Bulletin 180.* Washington: Government Printing Office, 1961, pp. 173-95.

Bibliography

La Barre, Weston. *The Ghost Dance: Origins of Religion.* London: George Allen & Unwin, 1970.

Lame Deer, John Fire. *Lame Deer, Seeker of Visions: The Life of a Sioux Medicine Man.* Ed. by Richard Erdoes. New York: Simon & Schuster, 1972.

MacLeod, William Christie. "Conservation among Primitive Hunting Peoples," *Scientific Monthly*, vol. 43 (July-Dec. 1936), pp. 562-66.

Mails, Thomas E. *Dog Soldiers, Bear Men, and Buffalo Women: A Study of the Societies and Cults of the Plains Indians.* Englewood Cliffs, N.J.: Prentice-Hall, 1973.

Martin, Calvin. "The European Impact on the Culture of a Northeastern Algonquian Tribe: An Ecological Interpretation," *William and Mary Quarterly*, vol. 31 (January 1974), pp. 3-26.

———. "Fire and Forest Structure in the Aboriginal Eastern Forest," *The Indian Historian*, vol. 6, no. 4 (Fall, 1973), pp. 38-42, 54.

———. *Keepers of the Game: Indian-Animal Relationships and the Fur Trade.* Berkeley: University of California Press, 1978.

Momaday, N. Scott. *The Way to Rainy Mountain.* Albuquerque, New Mexico: University of New Mexico Press. 1969.

Mooney, James. "The Ghost-dance Religion and the Sioux Outbreak of 1890," *Fourteenth Annual Report of the Bureau of Ethnology* to the Secretary of the Smithsonian Institution, 1892-1893, by J. W. Powell, Director. Washington: Government Printing Office, 1896. pp. 641-1110.

Morey, Sylvester M. *Can the Red Man Help the White Man? Denver Conference with the Indian Elders, 1968.* New York: G. Church, 1970.

———, and Olivia L. Gilliam, eds. *Respect for Life: The Traditional Upbringing of American Indian Children.* Garden City, New York: Waldorf Press, 1974.

Nash, Roderick, ed. *Environment and Americans: The Problem of Priorities.* New York: Holt, Rinehart & Winston, 1972.

Norton, Jack. "To Walk the Earth," *The Indian Historian*, vol. 7, no. 4 (Fall, 1974), pp. 27-30.

Ortiz, Alfonso, ed. *New Perspectives on the Pueblos.* Albuquerque: University of New Mexico Press, 1972.

———. *The World of the Tewa Indians.* Chicago: University of Chicago Press, 1963.

Paulsen, Ivar. "The Animal Guardian: A Critical and Synthetic Review," *History of Religions*, vol. 3, no. 2, 1964.

Radin, Paul. *The World of Primitive Man.* New York: E. P. Dutton, 1971.

Reed, Gerald. "A Native American Environmental Ethic: A Homily on Black Elk," Unpublished MS, Mid-America Nazarene College, Olathe, Kansas.

Reichard, Gladys A. *Navajo Religion.* 2 vols. New York: Pantheon Books, 1950. (Reprint in one volume, Princeton: Princeton University Press, 1963).

Richardson, Boyce. *Strangers Devour the Land.* New York: Alfred A. Knopf, 1975.

Ridington, Robin and Tonia Ridington. "The Inner Eye of Shamanism and Totemism," *History of Religions* 10 (1970):49-61.

Ritchie, William A. "The Indian in His Environment," *The New York State Conservationist*, (Dec.-Jan. 1955-56), pp. 23-27.

Roe, Frank Gilbert. *The Indian and the Horse.* Norman, Okla.: University of Oklahoma Press, 1955.

Bibliography

———. *The North American Buffalo: A Critical Study of the Species in its Wild State.* Toronto: University of Toronto, (1951) 1970.

Roedinger, Virginia. *Ceremonial Costumes of the Pueblo Indians.* Berkeley: University Press, 1961.

Safford, William E. "Our Heritage from the American Indians," *Annual Report of the Board of Regents of the Smithsonian Institution,* 1926. Washington: Government Printing Office, 1927, pp. 405-10.

Sauer, Carl Ortwin. *American Agricultural Origins: A Consideration of Nature and Culture,* ed. by Robert H. Lowie. Berkeley: University of California Press, 1936.

Schultz, Robert C., and J. Donald Hughes, eds. *Ecological Consciousness.* Washington, D.C.: University Press of America, 1981. Note particularly "The American Indian as Miscast Ecologist" by Calvin Martin, pp. 137-48.

Scully, Vincent. *Pueblo: Mountain, Village, Dance.* New York: Viking Press, 1975.

Sinyella, Juan. "Havasupai Traditions," edited and with an introduction by J. Donald Hughes, *Southwest Folklore* 1 (Spring 1977):35-52.

Smith, David Merrill. *Inkonze: Magico-Religious Beliefs of Contact-Traditional Chipewyan Trading at Fort Resolution, N.W.T., Canada.* Ottawa: National Museum of Man, Ethnology Division, Paper No. 6, Mercury Series, 1973.

Snow, John. *These Mountains are Our Sacred Places.* Toronto: Samuel-Stevens, 1977.

Speck, Frank G. "Aboriginal Conservators," *Bird Lore,* vol. 40, 1938-39, pp. 258-261.

———. *Family Hunting Territories and Social Life of Various Algonkian Bands of the Ottawa Valley.* Ottawa: Government Printing Bureau, 1915. (Canada Department of Mines, Geological Survey, *Memoir 70,* no. 8, Anthropological Series).

———. "The Indians and Game Preservation," *The Red Man,* vol. 6, no. 1, Sept. 1913, pp. 21-25.

———. "Savage Savers," *Frontiers,* vol. 4, no. 1, Oct. 1939, pp. 23-27.

Spradley, James P. *Guests Never Leave Hungry: The Autobiography of James Sewid, A Kwakiutl Indian.* New Haven: Yale University Press, 1969.

Steward, Julian H. "Ecological Aspects of Southwestern Society," *Anthropos* 32 (1937), pp. 87-104.

Stewart, Guy R. "Conservation in Pueblo Agriculture," *Scientific Monthly,* vol. 51, no. 4 (1940), pp. 201-20, 329-40.

———, and Maurice Donnelly. "Soil and Water Economy in the Pueblo Southwest," *Scientific Monthly* 56 (1943), pp. 31-44, 134-44.

Tanner, Adrian. *Bringing Home Animals: Religious Ideology and Mode of Production of the Mistassini Cree Hunters.* Social and Economic Studies No. 23, Institute of Social and Economic Research, Memorial University of Newfoundland, 1979.

Tedlock, Dennis, and Barbara Tedlock, eds. *Teachings from the American Earth: Indian Religion and Philosophy.* New York: Liveright, 1975.

Terrell, John Upton and Donna M. Terrell. *Indian Women of the Western Morning.* New York: Dial, 1974.

Tyler, Hamilton A. *Pueblo Animals and Myths.* Norman, Oklahoma: University of Oklahoma Press, 1975.

———. *Pueblo Gods and Myths.* Norman, Oklahoma: University of Oklahoma Press, 1964.

Tyler, S. Lyman. *A History of Indian Policy.* U.S. Department of the Interior, Bureau of Indian Affairs, 1973.

Bibliography

Udall, Stewart L. "The Indians: First Americans, First Ecologists," in Iacopi, Robert L., Bernard L. Fontana and Charles Jones, eds. *Look to the Mountain Top*. San Jose, California: Gousha, 1972.

———. "The Land Wisdom of the Indians," Chapter 1, *The Quiet Crisis*. New York: Holt, Rinehart & Winston, 1963.

Underhill, Ruth M. *Ceremonial Patterns in the Greater Southwest*. American Ethnological Society *Monographs* 13 (1948).

———. *Indians of the Pacific Northwest*. U.S. Department of the Interior, Bureau of Indian Affairs, Branch of Education, 1945.

———. *The Navajos*. Norman, Oklahoma: University of Oklahoma Press, 1956.

———. *Papago Indian Religion*. New York: Columbia University Press, 1946.

———. *Red Man's America: A History of Indians in the United States*. Chicago: University of Chicago Press, 1953.

———. *Red Man's Religion: Beliefs and Practices of the Indians North of Mexico*. Chicago: University of Chicago Press, 1965.

———. *Singing for Power: The Song Magic of the Papago Indians of Southern Arizona*. Berkeley: University of California Press, 1938.

Vecsey, Christopher, and Robert W. Venables, eds. *American Indian Environments: Ecological Issues in Native American History*. Syracuse, N.Y.: Syracuse University Press, 1980.

Vogel, Virgil J. *American Indian Medicine*. Norman, Oklahoma: University of Oklahoma Press, 1970.

Warren, Dave. "Man and His Environment: The American Indian and the Natural World," in *Environmental Awareness for Indian Education, Curriculum Bulletin No. 8* (1970), U.S. Department of the Interior, Bureau of Indian Affairs.

Wedel, Waldo R. "Some Aspects of Human Ecology in the Central Plains," *American Anthropologist*, vol. 55 (1953), pp. 499-514.

Whalen, Sue. "The Nez Perces' Relationship to Their Land," *Indian Historian*, vol. 4, no. 3 (Fall, 1971), pp. 30-33.

Wheat, Joe Ben. "A Paleo-Indian Bison Kill," *Scientific American* 216 (January 1967), pp. 44-52.

Whiting, Alfred F., et. al. *Hopi Indian Agriculture and Food*. Flagstaff, Ariz.: Museum of Northern Arizona, 1954.

Whorf, Benjamin Lee. *Language, Thought, and Reality*. Cambridge, Mass.: M.I.T. Press, 1956.

Wissler, Clark. "The Psychological Aspects of the Culture-Environment Relation," *American Anthropologist* 14 (1912):217-25.

———. *The Relation of Nature to Man in Aboriginal America*. New York: Oxford University Press, 1926.

———. "The Relation of Nature to Man as Illustrated by the North American Indian," *Ecology*, vol. 5, no. 4 (Oct. 1924), pp. 311-18.

———, ed. *Sun Dance of the Plains Indians*. New York: American Museum of Natural History, *Anthropological Papers*, vol. 16 (1921).

Witthoft, John. "The American Indian — Hunter," *Pennsylvania Game News*, vol. 29 (1953) Feb.: pp. 12-16; Mar.: pp. 16-22; Apr.: pp. 8-13.

Yazzie, Ethelou, ed. *Navajo History*, vol. 1. Many Farms, Arizona: Navajo Community College Press, 1971.

INDEX

A-badt-dadt-deah, 19
abalone, 44
acculturation, 119, *see also* assimilation
Acoma, 58, 71
acorn, 46, 49
Adams, Mount, 132
adaptation, 108
adobe, 100
adze, 53
aerial photographs, 71
agriculture, 4-7, 21, 23, 36, 51, 65-77, 99, 100, 103, 108, 116, 119-20, 124, 129, 139
air, 64, 86, 101, 106-107, 132-34, 138, 140, 142
airplane, 131
Alaska, 53, 131, 133-34, 136, 138
Alaska Native Brotherhood, 130
Alaska Native Settlement Act, 133
alcohol, 115, 119, *see also* wine
alder, 53
Algonquian, 6, 18-19, 66
Allegheny River, 128
allegory, 51
alluvial, alluvium, 70, 101
allotment, 124-25
Alquntam, 19
altar, 60, 74
amole, 51
ancestors, 63, 87, 107, 111, 135-36
Ancoutahan, 138
animals, 23-48 *and throughout*
animal dance, 51
Ani Yonwiyah, 60
ant, 42
antelope, 6, 24, 33, 36, 129, 142
anthropology, 4, 41, 95, 140
Apache, 12, 15, 19-20, 38, 42, 51, 68, 74, 102, 131
apology, to animals, 23, 28, 30, 33, 35, to plants, 50, to Mother Earth, 51-52, to trees, 54-55
April, 75
Arapooish, 142
Arbor Day, 141
arctic, 1, 8
Arikara, 68
Arizona, 60, 69, 141
Army, Secretary of the, 110
Army, United States, 112, 117, 123

arroyo, 101-102, 126
art, 10, 29, 65, 79, 93-94, 126
ash, tree, 2
Asia, 54
aspen, 51, 72
assimilation, 117, 124, 131, *see also* acculturation
Assiniboin, 39, 43, 85
Atlantic, 105
atoms, 16
autumn, fall, 35, 66, 74, 142
Awanichi, 140
axe, 2, 53, 98
Aztec, 75, 103

bacteria, 65
badger, 33, 35, 82
Badger Clan, 35
balance, of nature, 5-6, 15, 17-18, 25, 42, 59, 79, 90, 103
bark, 49, 52, 54-55, 85
basket, 10, 69-70
Bat Cave, 68
beach, 62
beads, 58
beans, 6, 65-67, 69, 72-73
bear, 16, 23-24, 26-28, 30-31, 33, 35, 42, 84-86, 92-93, 115, 129, 142
Bear Butte, 59
Bear Ceremony, 35
Bear Society, 85
Bear Tepee, 59
Beast Gods, 21, 28, 43
beauty, 13, 30, 50, 61, 107, 143
beaver, 24, 26-27, 37, 53, 93, 114, 142
Beaver Men, 30
bee plant, 69
bees, 4
Bella Coola, 19, 47, 51, 54
berries, 46, 49, 62, 85
Betatakin, 101
Big Snake Way, 27
biodegradable, 100, 135
biosphere, 22, 140
birch, 6, 52
birds, 2, 6, 10, 12, 14-16, 23-24, 33, 35, 37, 43, 45-46, 55, 59, 67, 73, 76, 78-81, 83, 86, 89, 91, 93, 98, 106, 114, 122, 131
birth, 95

162

Index

bison, *see* buffalo
Black Elk, 14, 17-20, 44, 67, 84
Blackfeet, 30, 108,109, 114
Black Hawk, 19
Black Hills, 119
Black Mountain, 63
blanket, 20, 47, 115-16
blood payment, 66
bluebird, 33
bluejay, 12, 16, 20
Blue Lake, 60, 132
boats, 8
bobcat, 33
bows and arrows, 38, 98
Brave Buffalo, 80
bread, 52
Britain, British, 111, 114-15
British Columbia, 23, 130
Brown, Joseph Epes, 15
brush, 71
Bryan, Kirk, 70
buffalo, 1-2, 4, 6-7, 25, 29-31, 38-43, 56, 67-68, 79, 83, 89-90, 98, 106, 113-14, 119, 121-22, 131, 142, near-extinction, 112-13
buffalo calf, 89
Buffalo Dance, 33
buffalo drives, 40-41
Buffalo Society, 85
bulbs, 49
bullhead, 47
bull-roarer, 76, 93
bundle, sacred, 28, 89
Bureau of American Ethnology, 138
Bureau of Indian Affairs, 117, 125-27
Bureau of Land Management, 132
burning, 66, 72, 112
butchering units, 40
Butterfly House, 60

Cabeza de Vaca, 41
cactus, 21, 51, 75, 82, 100, 123
Cahokia, 99
calendar, 8, 72, 79, 90
California cultural area, 8, 34, 51, 96, 132
California, Gulf of, 52
campfire, 78, 138
Canada, 27, 49
canals, 71
candlefish, *see* eulachon
canoe, 6, 8, 22, 32, 52, 54, 87, 93, 115, 138

canyon, 60, 103, 134
Canyon de Chelly, 72
caribou, 37
carpentry, 53-55
carrying capacity, 95
Carson, Kit, 112, 127
Casper, Wyoming, 41
cattail, 74
cattle, 42, 102, 126
cave, 58, 82, 100
cedar, 7, 46-47, 53-55, 86, 123, 132
Celilo Falls, 128
census, of animals, 37-38, 79
center, 59
Central America, 75, 99
ceremonies, 11, 14-16, 21, 23-24, 28, 31, 33, 35, 43, 46, 50-53, 59, 65-66, 74, 76, 79, 84-85, 88-92, 97, 103, 106, 116, 123, 141, *see also* ritual
Chac, 75
Chaco Canyon, 70, 101-102
Champlain, Lake, 56
change, 139, 143
Changing Woman, 58, 82
chants, 7, 43, 50, 74, 138, *see also* songs
cheese, 54
Cherokee, 19, 60, 95, 111-12, 128-29
cherries, wild, 41, 72
chestnut, 53
Cheyenne, 17-18, 29-30, 39, 50-51, 89, 97, 119
chicken, 37
children, 21, 24-25, 32, 34, 43, 45, 60, 63, 65-66, 73-74, 97, 106
chili, 108
Chilkat, 138
China, 115
Chinle, 63
Chi'óol'íí, 58
chisel, 53
Choctaw, 38
Chumash, 44, 51
cicada, 82
circle, of life, sacred, 20, 22, 59-60, 65, 107
citizens, 125, 130
Civilian Conservation Corps, 126
civilized tribes, 112
Civil War, 63, 115
clans, 5, 35, 61, 87, 95
clay, 7, 51
Clayoquot, 22
Clear Lake, 34

163

Index

climate, 57, *see also* weather
climatic change, 101-103
clouds, 10, 13, 16, 53, 73-76, 91
clown, 90
Clutesi, George, 81
coal, 52, 134, 141
coffee tree, 53
cold front, 16
Collier, John, 125-26
color, 59, 83
Colorado, 40-41
Colorado, University of, Museum, 40
Colorado Desert, 1
Colorado River, 70, 94
Colorado Rockies, 56
Columbia River, 120, 128
Comanche, 44
comic, 90
Commissioner of Indian Affairs, 113, 125
community, 16, 25, 76, 118
competition, 61
Congress, 60, 110, 124-25, 132
conservation, 14, 34, 36-37, 44-45, 47, 50-52, 56, 62, 68, 102, 108, 112, 116, 125-26, 130-31, 135, 137, 139
conservation movement, 133
constitution, tribal, 112
Constitution, United States, 110
contraception, 98
cooperation, 61
copper, 47, 52, 99, 115
corn, 1, 4, 6, 8, 16-17, 31, 35, 65-66, 68, 71-74, 83, *see also* maize
corn, Hopi, 69
Corn Maidens, 21
cornmeal, 52, 61, 74, 90
Corn Mother, 19
Corn Priest, 68
Coronado, 5
cosmology, 59
costumes, 10, 50, 69, 88, 90, 92
cotton, 69
cottonwood, 2, 7, 21, 44, 53, 72, 90, 142
courts, 63, 112, 131
coyote, 12, 20, 33, 35, 82
creation, 23, 46, 83
Creator, 2, 19-21
crow, bird, 79, 84-85, 122
Crow Indians, 18-19, 44, 142
culture hero, 21
cycles, of nature, 80, 87, 89, 94-95, 140, 143

Da, Popovi, 9, 11
Dakota, 5, 18, 31, 48, 80, *see also* Lakota, Sioux, Teton
dams, 70-71, 102, 125, 128-29
dance, 7, 14, 16, 24, 31, 33, 46, 51, 53-54, 66, 73, 86, 89-90, 92-93, 100, 121-23, 143
Dark Dance, 35
dawn, 13
death, 19-20, 24-25, 28, 55, 74-75, 78, 95-96, 121, 122
deer, 1, 4, 6, 17, 24, 27, 30, 32, 34-38, 42, 46, 55, 76, 82, 84, 99, 106, 113-14, 122, 129, 131, 138, 142
December, 36
Deer Dance, 33
Deer Owner, 33
deforestation, 57, 64, 102
deist, 4
Delaware, 35
Deloria, Vine, Jr., 136, 139
democracy, 17
demographers, 95-96
dentalium shells, 47
Denver, 64
Department of Conservation, 103
desert, 1-2, 69, 82, 94, 103
devil's claw, 69
digging stick, 20, 67, 69-70, 73, 106, 118-19
Dineh, 5
Diogenes, 48
Dionysiac, 92
directions, 15, 33, 59, 75, 89
disease, 3, 21, 40, 43, 96, 107, 115-16, *see also* illness, sickness
ditches, 70-71
dog, 31, 43-45, 98
Dog Soldiers, 39
dominion, 17
Douglas fir, 53
dove, white-winged, 72
drama, 92
dream, 27, 35, 80, 83, 87-88, 120
drought, 6, 67, 70, 101-103
drum, 76, 92
duck, 34
Duncan, William, 118
Durham, Jimmie, 60, 128
dust bowl, 106, 134
Duwamish, 11

164

eagle, 10, 13, 15, 21, 24, 28, 31-33, 35-36, 87, 90, 93, 122, 131
Eagle Dance, 33
earth, 3, 6, 7, 11-15, 20, 22, 25, 38, 48, 50-52, 59, 62, 65-67, 71, 80, 82-84, 89-91, 95, 99-101, 103-105, 120-22, 125, 133, 135-37, 139-41
earth, sacred, 11, 14
Earth Day, 141
Earth Goddess, 20
Earthmaker, 19
earth renewal, 121
Eastman, Charles, 48, 80, 142-43
ecologists, 22, 55, 95, 137
ecology, dynamic, 139
economic, economy, 5, 14, 105, 119, 131, 134
ecosystem, 50, 80, 96, 111, 114, 139-40
Eden, 2
education, 9, 117-18, 135
elderberry, 46
Elder Brother, 21
elders, 6, 20, 27, 50, 78-94
elk, 6, 16, 36, 53, 78, 122, 131, 142
elk-dog, 43, 108
Elk Society, 30
elm, 2, 6, 52
Eloheh, 60
emergence, 52
energy, 22, 97, 136, 141
energy development, 134
English, 2, 109
ephedrine, 98
epidemiology, 96
erosion, 44, 66, 69-71, 76, 101-102, 106, 116, 132
Esaugetuh Emissee, 19
Eskimo, 8
ethics, environmental, 14, 18, 47, 61, 81, 98, *and throughout*
ethnic science, 79, 82, 140
eulachon, 8, 46, 115
Europe, 66, 114-15
European, 1-5, 7, 14, 16-17, 54, 56, 62-63, 69, 81, 96, 99-101, 105-110, 118, 137
Everglades, 8
exhaustion of resources, 99-100
explanation, 16
exploitation of earth, 95, 135
explorers, 3, 99, 137
extermination, 112, 117
extinction, 81, 112-13

factory system, 110
family, 5, 61-62, 95, 98, 118-19
family hunting ground, 37
farmers, farming, farms, 3, 20-21, 37, 65-77, 106, 118-19, 124, *see also* agriculture
fasting, 27, 84, 86-87
Father Sun, 91
feathers, 12, 15, 28, 31, 32, 36, 46, 52-53, 74, 90-91
feedback, 5
female, feminine, 15, 19, animals, 33-35
fertility, fertilization, fertilizing, 20, 65-67, 72, 90
fiber, 51
finch, yellow, 12
fir, 7, 53
fire, 4, 6, 21, 28, 31, 38, 43, 53-54, 66, 125-26, *see also* burning
fire drives, 55
firefighting, 125
fires, forest, 55-57
firewood, 5, 7, 40, 52-53, 102
First Salmon Ceremony, 46
fish, fisheries, fishing, 2, 3, 6, 8, 16, 20, 23, 32, 37, 44-47, 54, 62-63, 66, 83, 86, 98-99, 106, 110, 115-17, 120-21, 125-31, 134
fishing, commercial, 116, 130
Fish-In, 130
Flathead Indian Reservation, 113
flint, 20, 99
flood, 6, 90, 102
floodwater irrigation, 70, 72, 101
flowers, 6, 15, 50-51, 79, 138
folklore, 81, *see also* myths, stories
food, 7, 23, 25, 27, 30, 39, 46-47, 49-51, 61, 68, 76, 79, 81, 83, 90, 92, 95, 100, 102-103
food chain, 79, 83
forest, 1-2, 4, 6, 8, 12, 17, 21, 23, 27, 34, 52-53, 55, 57, 62, 66, 78-79, 98-99, 102, 107, 115-17, 124-25, 127, 132, 135
foresters, 56
Forest Service, 127
Fort Berthold, 128
Fort Simpson, 118
Fort Sumner, 63, 112
fox, animal, 84
Fox Indians, 53
France, French, 109, 111

165

Index

Francis, Saint, 48
Frison, George C., 41
frog, 43
frost, 67, 69, 72
fruit, 20, 50-51, 53, 66, 79, 83, 108, 112, 121
fur, 23, 107, 109, 114, 116

game park, 131
games, 21
Garden of Eden, 2
gas, natural, 125, 134
gathering, 6, 16, 23, 29, 49-51, 103, 117
geese, 79
General Allotment Act, 124
geography, sacred, 58-59
geology, 101
Georgia, 112
germs, 16
Ghost Dance, 121-23
giants, 21
gift, 15, 19, 21, 26, 46-48, 50, 58, 61-62, 83, 86-87, 89, 115
gift to the place, 58
Gila River, 71
Giver of Breath, 19
glacier, 87, 138
goat, 44, 102
Gobernador Knob, 58
gods, 13, 18, 21, 30, 33, 43-44, 64, 91
gold, 112
Goldman, Irving, 25
Good Hunter, 26
Gosiute, 36
Gossypium hopi, 69
gourd, 69, 76
grain, 61
Grand Canyon, 8, 52, 102, 108, 132, 142
Grand Coulee Dam, 128
Grandmother Earth, 20
grass, grassland, 1, 6, 11, 15, 44, 51, 56, 58, 80, 89, 106, 121, 138, 142
grasshopper, 13, 74
gray jay, 34
Great Basin, 8
Great Drought, 101
Great Elk, 78, 83
Great He-She, 19
Great Keeper, 33
Great Lakes, 111
Great Mystery, 3, 18

Great Plains cultural area, *see* Plains
Great Spirit, 1, 18-19, 22, 52, 60, 63, 105, 109, 122
Great White Bear, 34
grizzly bear, 92
grouse, 36
guardian, 16, 27, 31, 78, 84, 86-87, 92, 94
Gulf of California, 52
Gulf of Mexico, 99
guns, 98, 108, 112, 115, *see also* rifles

Haida, 19, 32, 54, 92, 132
halibut, 8
hammer, 53
harmony, 6, 13-16, 48, 64, 72, 74, 107, 122, 137, 139, 141-43
Harrison, Benjamin, 113
harvest, 73-74
harvest dance, 73
Havasupai, 52, 58, 68, 70-71, 74, 108, 132-133
hawk, 93
healing, 16, 42-43, 74, 86-88
healing societies, 43
Heamma Wihio, 18
Heizer, Robert F., 139
hemlock, 7
Hennepin, Louis, 39
herb, 51, 85
herring, 8
hickory, 2
Hidatsa, 18, 44, 84
hides, 7, 28
Higheagle, 119
historians, 140
history, sacred, 58, 68, 81, 83
hoe, 67, 69, 106
hogan, 59
Hohokam, 71
ho'i, 15
holy occupation, 25, 65
Homer, 12
Homeric simile, 87
Homestead Act, 124
hoochinoo, 119
Hopi, 14-15, 18, 21, 33, 35-36, 43, 52-53, 59-60, 63, 65, 68, 72-75, 79, 91, 100, 102-103, 120, 142
horn, 53
horse, 7, 38-44, 67, 98, 102, 108, 114, 120, 142
Horseshoe Mesa, 52

166

Index

house, 6, 8, 13, 45, 52, 54, 60-61, 86, 100, 115, 142
House of Snow, 60
House of Wolves, 92
hozho, 13
Hudson's Bay Company, 114
hunting, 3-7, 15, 16, 21, 23-49, 52, 61-63, 66-67, 78-79, 82, 85, 87-88, 97-99, 103, 110, 118-20, 127, 129, 131, 139, 143
hunting, commercial, 114-15
hunting grounds, territory, 37-38, 42, 60, 117

Illinois, 99
illness, 85, *see also* disease, sickness
Inca, 103-104
Indian Claims Commission, 127
Indian Reorganization Act, 125-26
Indian Territory, 117
individual, 16, 61
initiation, 60
injury to nature, avoidance of, 17-18, 20, 25, 38, 50, 52-53, 55
insects, 10, 23, 57, 67, 91
instrumental value, 140
intellectual curiosity, 16
Interior, Department of the, 117, 125
Interior, Secretary of the, 137
International Treaty Organization, 128
Intiwa, 15
Iowa, 69
iron, 115
Iroquoian, 6
Iroquois, 21, 35, 62, 66, 98-99
irrigation, 70-72, 74, 126, 129
Isleta Pueblo, 14, 35, 59
Itiwana, 59
Iyatiku, 21

Jacobs, Wilbur, 137
jadeite, 53
jay, 12, 16, 20, 34
Jemez Pueblo, 120
jerky, 41
Jicarilla Apache, 131
jokes, 19
Jones, "Buffalo" (Charles Jesse), 40
Jones, Edwin, 41
Joseph, Chief, 63, 105
June, 72
juniper, 5, 101-102
Justice, Department of, 131

kachina, 33, 53, 60, 90-91, 94
Kadachan, 138
Keeper, 32-33
Keres, 21
Kicking Bear, 122
kidney beans, 69
kinship, 6, 17, 25-26, 35, 62
Kinzua Dam, 128
Kiowa, 58, 122
kiva, 59-60, 74, 100-101
Klamath, 127
knife, 53, 67, 85, 115, 121
Knox, John, 110
Kwakiutl, 5, 22, 26, 43, 46, 51, 55, 92

lake, 24, 58, 60, 79, 129, 131, sacred, 36, salt, 52
Lake of Whispering Waters, 60
Lakota, 15, 24, 136, *see also* Dakota, Sioux, Teton
land, 20, 42, 60-64, 78, 81, 86, 105-136, 138-39, 141, public, 117, 132-33
landscape, 59, 100, 141
languages, 5-6, 14, 19, 21, 24, 54, 66, 112, 119, 138
lark, 22
lark, horned, 6
lark bunting, 6
lark sparrow, 6
Larocque, 44
lava, 58
leather, 6, 58
leaves, 52, 79-80
legume, 65
Leschi, Chief, 111
Life Giver, 19
lightning, 6, 13, 21, 76
lima beans, 69
linguistics, 4
liquor, 113
Little Deer, 32
llama, 98
locality, 58-59, 63, 74, 80
lodges, 6, 10, 44, 90
logging, 116, 132
logs, 71
longhouse, 35, 98
Long-Life-Maker, 55
Los Angeles, 64, 141
love, of earth, 14, 62, of nature, 10-14, 138

167

macrocosm, 89
Maine Woods, The, 137
maize, 7, 21, 23, 65-67, 71, 73-74, 119, *see also* corn
Makah, 32, 47
Makers, 19
male, masculine, 15, 19, animals, 33, 35
mandala, 10
Mandan, 20, 68, 84, 89
Manitou, 18-19
Many Tail Feathers, 40
maple, 2
Maquinna, 32
Marcy, R. B., 39
marmot, 35
marriage, 24, 97
marshes, 6
marten, 115
Masau'u, 21
mask, 8, 21, 24, 43, 54, 86, 90-92
Massachusetts Indians, 66
Massaum, 51
Master of Animals, 32, 34
Master of Breath, 21
Master of Game, 33
Master of Life, 18, 106
mato tipi, 59
maxpe, 18
Maya, 75, 103
measles, 96
medicine, 49-51, 79, 85, 88, 98
Medicine Creek, 110
medicine dance, 143
medicine men and women, 16, 29, 42-44, 49, 87-88
Meeker, Ezra, 111
melons, 73
men, 28, 49, 67, 69, 74, 106, 118
menhaden, 66
Menominee, 51, 127
menstrual cycle, 98
mesa, 60, 70, 103, 134
Mesa Verde, 70, 101-102
mesquite, 72
metate, 52
meteor shower, 109
Metlakatla, 118, 133
Mexico, 54, 75, 99, 111, 123
mica, 99
Micmac, 26, 31
microclimates, 82
microcosm, 10, 89
minerals, 51-52, 59, 126, 131

miners, mines, mining, 20, 109, 112, 116-17, 132
Minnesota, 131
mint, 69
Mississippi River, 111
Mississippi Valley, 99
Missouri River, 119
Mixtec, 103
moccasins, 7, 34
mockingbird, 21
Moenkopi, 101
mole, 33
Momaday, N. Scott, 59
Monk's Mound, 99
monotheism, 108
monster, 58, 86
Montanists, 48
Montezuma, 21
moon, 10-11, 21, 47, 72, 75
Mooney, James, 114
moons, 79
moose, 4, 26, 34, 37
moosefly, 32
morache, 76
Morgan, Thomas J., 113
Morning Star, 21
mortars, 66
Mother Earth, 1, 14, 20, 50-52, 61, 65-67, 71, 82, 83, 89, 95, 103, 119-21, 141
Mother Nature, 51, 142
Mother of the Beasts, 33
Mound Builders, 99-101
mountain goat, 24, 28, 43
mountain lion, 24, 33, 82
mountains, 3, 8, 10-12, 15-16, 20, 22, 24, 32, 45, 53, 58-60, 70, 86, 105, 113, 117, 120, 138, 142, sacred, 63, 100
mountain sheep, 35, 129, 142
mouse, 73
Muingwu, 21
Muir, John, 138-39, 142
mulberry, 53
multiple use, 133
music, 92, *see* song
mythology, myths, 6, 19, 23-24, 33, 52, 58, 81-83, 86

names, 27, 60, 62, 74, 79, 84, 86-87, avoided, 28
Nash, Roderick, 98
Natchez, 99

Index

National Bison Range, 113
National Forests, 124, 126-27, 132-33
National Indian Youth Council, 131
National Parks, 124, 132-33, 141
Native American Church, 123
nature, appreciation of, 10-14, control of, 17, law of, 121, love of, 10, order of, 111
Navajo, 5, 12-13, 20-21, 23-24, 27, 30, 33, 35-36, 42-44, 50, 53, 55, 58-59, 63-65, 68, 70, 72, 74-75, 79, 82, 102, 112, 126, 141, 143
nest, 15
Nevada, 36, 121, 129
New Life Lodge, 89
New Mexico, 9, 58, 68, 100, 112, 141
newspaper, 112
Nez Percé, 105
night, 20,93
night chant, 13
nitrogen-fixing, 65, 72
nomads, 6, 114
Nootka, 31-32, 92-93, 115
North Carolina, 99
North Dakota, 128
Northeast Forest cultural area, 6, 34, 37, 65-66, 128
Northwest Coast cultural area, 7-8, 10, 24, 27, 43, 45-47, 53-55, 62, 85-87, 92-94, 97, 114-16, 119, 128, 130-32, 134
North West Company, 114
North Wind, 51
nuts, 49, 53
Nuvatukya'ovi, 60

offerings, 15, 25, 27-30, 33, 43, 50, 58, 72, 74, 76, 84, 90
Ohio Valley, 111
Ohiyesa, 48, 80, 142-43
oil, 8, 54, 125, 133-34
Ojibwa, 37, 131
Oklahoma, 111, 117
Olmec, 103
Omaha, 2, 68, 83
oneness, 14-15, 82, 88
Onondagas, 99
Oraibi, 73, 101-102
orca, 8, 25, 86, *see also* whale
Oregon, 78
orenda, 18
origin, 82, *see also* creation
Osage, 10, 29, 83

Osheshewakwasinowinini, 37
otter, 35, 115, *see also* sea otter
Overlord of Fish, 32
owl, 24, 43, 79, 84-85
ownership, 3, 61-64, 106, 109-110, 124
oxygen cycle, 22, 81
Ozarks, 99

Pacific, 1, 7
paint, painting, 27-28, 54, 68, 74, 92
Paiute, 8, 121, 129, 137
Pan-Indian, 123
Papago, 12, 15, 19, 21, 28-30, 36, 42, 51-52, 65, 68-73, 75-76, 82, 90
partridge, 37
passenger pigeons, 2, 114
Paul, Aleck, 37
Pawnee, 10, 18, 40, 84, 116
Penobscot, 137
Peru, 103
petroglyphs, 10
Peyote, 123
philosophy, 4, 14, 22
Piegan, 40
Pierce, Franklin, 64
pig, 37
piki, 52
pilgrimage, salt, 52
Pilgrims, 66
Pima, 68-73, 108, 129
pine, 11, 53, 102
piñon, 53, 102
pipe, sacred, 15, 29, 67-68, 89
place, 58-60, 63, 65, 80, 141, sacred, 12, 63
Plains cultural area, 3, 6-7, 25, 29-31, 36, 43, 38-44, 50, 52-53, 56, 58-60, 66-68, 83, 89, 94, 98, 108, 111-14, 119, 120-21, 123, 134, 142
planters, 16, 78
planting, 72, 74, 88
plants, 49-57 *and throughout*
Plateau, 120
plaza, 59, 100
Pleiades, 72
plow, 2, 62, 67, 69, 98, 106, 108, 118-21, 134
plum, 53
poetry, 12, 137
Point Elliott, 110
Polike, 60
Polis, Joe, 137
pollen, 13, 50-51, 74, 101

169

pollution, 1, 45, 52, 64, 86, 101, 129, 132, 134
Pomo, 34
ponderosa pine, 102
population, 95-104, 112
porcupines, 24, 36
Poshaiangkaia, 21
potatoes, 69, 115
potlatch, 47, 115, 119
potters, pottery, 10, 51
Powell, John Wesley, 137-39
power, 17-18, 21, 23, 25, 27-28, 31-32, 35, 54, 56, 58, 60, 79-80, 83, 86-87, 89, 97, 105, 118, 129, 141, 143
Power of the Shining Heavens, 19
power saw, 98
prairies, 2, 117, *see also* grassland, plains
prairie dog, 73
prayer, 14, 20-22, 27, 29, 32-33, 50-53, 72-74, 90, 123, 136
prayer sticks (wands), 13, 28, 52, 74, 76, 90-91, 100
pre-Columbian, 4, 6, 9, 95-96, 98, 103, 123
predators, 14, 33-35
prey, 14, 36
Price, Pete, 65
prohibition, 119
property, *see* ownership
psychology, 4
puberty, 87
Pueblo, 7-9, 14, 16, 28, 30, 33, 42-43, 53, 59-61, 67-70, 82, 90, 94-95, 97, 100-103, 109, 118, 120, 132, 134, 142, abandonment, 101-102
Puget Sound, 111
pumpkin, 69
purification, 27-28, 31, 35, 85-87, 89, 116
Puyallup, 45
Pyramid Lake, 129
pyramids, 3, 99

quail, 36
quarry, rock, 52
Quillayute, 86
quinine, 98

rabbit, 17, 20, 31, 36
raccoon, 32
Radisson, Pierre Espirit, 39
rain, 7, 13, 15, 33, 53-54, 68, 70, 73-75, 90-91, 101-102, 143
Rain Being, 76

rainbow, 13, 21
rain dance, 16
rake, 67
ranches, 109
rational, rationalist, 4, 16
raven, 24, 86-87, 93
reciprocal, reciprocity, 15, 17, 25, 105, 140-41
recycling, 22, 81, 99, 135, 139
red cedar, 53
red deer, 37
Red Lake Reservation, 131
redwood, 53, 132
Ree, 68
Reichard, Gladys, 50
relatedness, 15
relatives, 4, 13, 17, 19, 28-29, 51, 60, 79
religion, 4, 14, 25, 76, 81, 98, 106, 118, 120, 123, 135, 138-39, freedom of, 123, 126
relocation, 127
Remaining Group, 127
renewal, 89
resonance, 16, 103, 140
respect, 17, 33, 38, 45, 50, 52, 80-81
reservations, 111, 113, 117, 119, 123-25, 127-28, 131, 133-34
reservoirs, 125, 128
reverence, for life, nature, 6, 18, 35, 102, 125, 137, 141, 143
rice, wild, 6, 51, 131
rifles, 7, 38-39, 98, *see also* guns
Rio Grande, 7, 95, 103, 123
ritual, 13, 25-27, 29, 31-32, 35, 45, 50-51, 62, 66, 73, 75, 86-87, 90, 100-101, 105, 129, *see also* ceremonies
rivers, 4, 7, 8, 11, 16, 19, 103, 105, 130, 142
rocks, 11, 15, 52, 58, 70-71, 93, 138, igneous, 53
Rocky Mountains, 37, 56
rodent, 42
Roe, Frank Gilbert, 39, 43, 108
Roosevelt, Franklin, 125
Roosevelt, Theodore, 124
roots, 20, 49-51, 53, 69, 76, 79, 91, 100-101
roundness, 59, *see also* circle
Russian, 114-15

Sacred Arrows, 29
sacrifice, 15, 87, 90

Index

saguaro cactus, 21
sale of land, 63, 106
salinization, 71-72
Salish, Coast, 33
salmon, 1-2, 8, 24-26, 45-46, 62, 97, 115-16, 128, 132, 138
salmonberries, 93, 119
salt, 52, 71-72
Salt Lady, 52
salt lake, 52
salt weed, 142
Sanaco, 44
Sand Creek, 20
sand dunes, 41
sand painting, 10, 43, 59
sandstorm, 73
San Francisco Peaks, 60
San Ildefonso Pueblo, 9
sanitation, 100
Savage, Major, 140
saw, 53, 98
schools, 112, 118-19, 135
Schweitzer, Albert, 18
science, 79-80, 96, ethnic, 79, 82, 140
Scully, Vincent, 103
sculpin, 86
sculpture, 92
sea, 7, 20, 22, 24-25, 45, 81, 86, 95, 111
sea lions, 8
seals, 8, 93, 115-16, 138
sea otters, 8, 114-15
season, hunting, 36
seasons, 21, 55, 72, 79, 82-83
Seattle, Chief, 11, 62, 64, 95, 111
Second World War, 127, 130
seeds, 53, 66, 69, 73-74, 82, selection, 68
Senate, 110
Seneca, 128
sex, 27, 49, 86-87, 90, 97
sexual abstinence, chastity, continence, 27, 87, 97
sgana, sganagua, sgaga, 32
Shakers, 48
shaman, 35, 119
sharing, 41-42, 47-48, 140
shark, 8, 54, 87, 93
sheep, 37, 44, 102
shell, 28, 53, 76, 95, 99
shellfish, 130
shield, 84-85, 89
Shiprock, New Mexico, 58

ships, 1, 114
shonka wakan, 43
Shooter, 119
Short Bull, 122
Shoshoni, 108-109
Shotuknangu, 21
shrew, 33
shrub, 53
Siberia, 53
sickness, 16, 22, 26, 30-32, 42, 87, see also disease, illness
Sierra Club, 138
singers, 21, see also song
single prey species, 36
Sins Sganagwai, 19
Siouan, 18
Sioux, 14, 15, 44, 59, 68, 85, 113, 119, 122-23, 131, 136, 142, see also Dakota, Lakota, Teton
Sioux, Teton, 11, 20, 80, 84, 119
Sioux City, 39
sipapuni, 59
Sitka, 115
Sitting Bull, 121
sixth sense, 78
sky, 10, 20-21, 54, 90
slash and burn, 66
sleds, 6
smallpox, 96, 109, 115-16
smelts, 8, 47
Smith, Henry, 11
Smohalla, 120, 134
smoke, 75, 101
smokehouse, 62, 115
Smokey the Bear, 56
snail darter, 128
snake, 16, 27, 33, 35
Snake-Antelope Ceremony, 33
snow, 7, 38, 73, 84, 142
snowshoes, 6
soap, 51
societies, 35, 85, 92
soil, 65-66, 68-73, 99, 106, 112, 116, 125, 138
Soil Conservation Service, 70
soil exhaustion, 99
soil moisture, 69
soil type, 57
song, 10-11, 14, 21, 24, 27, 29, 31-32, 43, 46, 50, 65-66, 68, 73-74, 84-86, 88-89, 92-93, 122-23
Song of the Hair Seals, 93

171

Index

Southeastern cultural area, 4, 8, 38, 67, 111
Southwest cultural area, 7, 33, 44, 64, 68-77, 90, 100-102, 119, 134
South Wind, 51
space, 63
Spanish, 64, 71, 100, 103, 108-109, 114, 119
sparrow, 6
Speck, Frank G., 37
spider, 18
Spider Old Woman, Spider Woman, 15, 21
spirit beings, 3-5, 10, 15-16, 18-19, 21, 23, 25, 27, 30, 32-33, 45, 49, 52-54, 58, 60, 75-76, 78, 83, 85, 87, 90-92, 94, 105, 129, 138
Spirit Lake, 127
Spirit of the Crops, 91
spring, season, 10, 51, 63, 67, 72, 82, 121
springs, 1, 7, 19, 58, 60, 70, 74, 101
spruce, 7, 24, 54, 59, 101, 132
Squanto, 66
squash, 6, 61, 66-67, 69
squirrel, 43, 67
Standing Bear, Luther, 3, 24, 80
Standing Rock Sioux, 131
stars, 10, 11, 15, 20-21, 72, 138
steward, 17
Stickeen, 138
stock reduction, 126
stones, 7, 20, 28, 58-59, 73, 80, 92, 98-100, 121
Stoney, 49
stories, 5, 78-94
streams, 7, 45, 53, 62, 80, 101, 132, 138, 142
strip mining, 134, 136, 141
subarctic forest, 56
succession, 57
summer, 13, 72, 74, 100, 132, 142
sun, 10-12, 18, 21-22, 54, 60, 63, 72-74, 81-82, 84, 87, 91, 100, 142
Sun Dance, 53, 89-90, 120, 123
sunflower, 67, 69
sunrise, 72
Survival of American Indians Association, 131
sweat bath, 83, 85-86, 101

Tahoe, Lake, 129
talisman, 28, 43, 59, 74, 84
Talking God, 21, 33
talking trees, 49, 52
Taos Pueblo, 15, 59-60, 63, 132
Tatanga Mani, 49, 52
Tatank Watcipi, 85
taxonomy, 79
tea, 51
teachers, teachings, 78-94, 110, 138
technology, 96, 98, 129, 136
Tecumseh, 63-64
Tellico Dam, 128
Temagami Band, 37
temples, 99
Tenaya, Chief, 140
tepary beans, 69
tepee, 6-7, 58, 84-85, 89, 94, 142
termination, 127
terraces, 71, 100, 102-103
Teton, 11, 20, 80, 84, 119, *see also* Sioux
Tewa, 15, 19, 79
thanks, 51-52, 69
Thanksgiving, 66
theater, 92, 100
thermodynamics, second law of, 16
Thoreau, Henry, 137, 139
thunder, 76, 102
thunder being, 79
thunderbird, 86, 90, 93
Thunder People, 21
tidelands, 129
timber, 99, 102, 124-25, 133
time, 63, 141
Tirawa, 18
Tlaloc, 75
Tlingit, 28, 35, 87, 115, 138
tobacco, 6, 28, 51, 53, 58, 66-67, 69, 74, 89, 115
toboggan, 52
Toelken, Barre, 120
Toltec, 103
tornado, 6
totem, 35
totem pole, 8, 54, 86, 118
tourists, 114
towns, 54, 59-60, 98, 100-101, *see also* villages
trade, 99, 103, 107, 109-110, 113-14
tradition, 8, 65, 81, 96, 106, 117-18, 126, 136-37, 140
trappers, traps, 36, 107, 114-15
trash disposal, 100-101
treaties, 63, 110-11, 119, 126, 130-31, 133

172

Index

tree rings, 101
trees, 3-5, 7, 11, 14-16, 21, 25-26, 49, 51-56, 60, 62, 64, 76, 79-80, 102, 108, 132, 138
tribal business, 126
tribal council, 131
tribal government, 126
tribal land, 42, 58-64, 117
tribal leaders, 66, 135
tribes, 5-7, 14, 16, 32, 47, 61, 86-87, 96, 108, 111, 127
Trickster, 19-20, 82
trout, 43
Tsimshian, 30, 36, 46, 118
tuberculosis, 96
turkey, 13
turquoise, 52
Turquoise Woman, 21
turtle, 36

Udall, Stewart, 137
Underhill, Ruth, 50, 75
underworld, 23-24, 52, 59, 82
United States, 63, 111-12, 114, 137, Indian policy, 109-136
universe, 7, 15, 18, 20, 22, 86, 89, 140
uranium, 134
urban life, 5, 99-100, 128
Ute, 19, 34, 37, 56, 102
utility, 105

Vecsey, Christopher, 107
vegetation, 51, 83, 91, 101, 143
villages, 3, 24, 27, 45, 47-48, 53, 55, 59-62, 88, 96, 99, 115, 117, 132-33, *see also* towns
visions, 40, 42-45, 60, 68, 83, 90, 122
vision quest, 27, 83-87
vow, 97

wakan, 18
Wakanda, Wakonda, 2, 18, 83
Wakantanka, 18-19, 80, 89
Walden, 137
Walking Buffalo, 49
Walking Coyote, 113
walnut, 2, 69
Walters, Winifred Fields, 136
War Department, 117
warfare, 26, 42, 62, 85-86, 96-97, ecological, 112
warrior, 88
warrior twins, 58

Warty, 20
Washington, D.C., 112, 121
Washington, state, 110, 129-30, 132
Washington Territory, 11
Washoe, 129
waste, avoidance of, 29, 31, 34, 37-42, 45, 50, 78, 81, 100, disposal, 135
water, 19, 22, 32, 45, 53, 58, 60, 68-71, 74-76, 78, 82, 85, 91, 93, 101-102, 106,107, 125, 128-29, 133-35, 138, 140
water cycle, 81
waterfall, 22, 138
waterfowl, 129
water monsters, 21
water serpent, 60
Wawatam, 31
Weasel Tail, 40
weather, 24, 26, 67, 73-74, 78-79, 82, 107
weaving, 10, 21, 92
weeds, 10, 67, 69, 73, 142
weirs, 45, 97-98
whales, 1, 8, 25, 32, 86-87, 93, 115
wheat, 108
Wheat, Joe Ben, 40
White Antelope, 20
White Buffalo Calf Women, 89
White Deerskin Dance, 92
Whiteshell Woman, 21
Wigaliva, 58
Wigwam, 31
wilderness, 2-4, 132-33, 138, 141
willow, 51
wind, 7, 21, 51, 68, 71, 73, 76, 95, 138
Windigo, 21
Wind River Valley, 142
wine, 21, 51, 75-76
winter, 6, 36, 38, 51, 72, 78, 92, 100, 142
Wisconsin, 127, 131
Wise One Above, 18
wolf, 25, 27, 33, 87, 138
Wolf-Chief, 44
Wolf Dance, 92
women, 46, 49, 53, 58, 66-67, 69, 74, 89, 106, 118-19
wood, 52-55, 58, 92, 106
woodcarvings, 8, 54
woodland, 6
woodworm, 23
Wounded Knee, 123

173

Index

Wovoka, 121-22, 134
Wyoming, University of, 41

Xupa, 18

Yakima, 132
Yé'iitsoh, 58
yellow cedar, 53
Yellow Moon, 75

yew, 53
yucca, 51, 72, 75, 101
Yuman, 68, 70
Yurok, 34, 36, 46, 92

Zapotec, 103
Zia, 76
Zuñi, 5, 12, 15, 19-21, 33, 35-36, 52-53, 58-60, 71-72, 74, 77